SUPERCONDUCTIVITY
A New Approach Based on the Bethe–Salpeter
Equation in the Mean-Field Approximation

SERIES ON DIRECTIONS IN CONDENSED MATTER PHYSICS

ISSN: 1793-1444

*For the complete list of the published titles in this series,
please visit: www.worldscientific.com/series/SDCMP

Series on Directions in Condensed Matter Physics – Volume 21

SUPERCONDUCTIVITY
A New Approach Based on the Bethe–Salpeter Equation in the Mean-Field Approximation

G P Malik
(Formerly) Jawaharlal Nehru University, India

World Scientific

NEW JERSEY • LONDON • SINGAPORE • BEIJING • SHANGHAI • HONG KONG • TAIPEI • CHENNAI • TOKYO

Published by

World Scientific Publishing Co. Pte. Ltd.

5 Toh Tuck Link, Singapore 596224

USA office: 27 Warren Street, Suite 401-402, Hackensack, NJ 07601

UK office: 57 Shelton Street, Covent Garden, London WC2H 9HE

Library of Congress Cataloging-in-Publication Data
Names: Malik, G. P., 1942–   author.
Title: Superconductivity : a new approach based on the Bethe–Salpeter equation in the mean-field
    approximation / G.P. Malik, (Formerly) Jawaharlal Nehru University, India.
Other titles: Series on directions in condensed matter physics ; vol. 21.
Description: Singapore ; Hackensack, NJ : World Scientific Publishing Co. Pte. Ltd, [2016] | 2016 |
    Series: Series on directions in condensed matter physics ; volume 21 |
    Includes bibliographical references and index.
Identifiers: LCCN 2015049520| ISBN 9789814733076 (hardcover ; alk. paper) |
    ISBN 9814733075 (hardcover ; alk. paper)
Subjects: LCSH: Superconductivity. | Bethe-Salpeter equation.
Classification: LCC QC611.92 .M35 2016 | DDC 537.6/23--dc23
LC record available at http://lccn.loc.gov/2015049520

British Library Cataloguing-in-Publication Data
A catalogue record for this book is available from the British Library.

The cover image is designed by Gautam Malik.

Printed in Singapore

The author wishes to dedicate this monograph to the memory of his father

Mr. R.K. Malik (March 22, 1914–August 8, 1996)

Preface

BCS theory (1957) emerged 46 years after the experimental discovery of superconductivity. This is a long period considering the large number of scientists — including some of the top names of the twentieth century — who toiled in this field. As is invariably the case for a new theory, it took time for BCS theory to be generally accepted — a status that it had by and large acquired prior to 1987. This however was the case when the highest T_c of any known superconductor (SC) was about 23 K. A radical change in the status quo occurred after Bednorz and Müller discovered an SC that had $T_c \approx 38$ K. An avalanche of activity followed thereafter — witnessed perhaps only once before when X-rays were discovered — leading to the discovery of several SCs such as YBCO and the Tl- and the Bi-based SCs that have T_cs greater than even the liquefaction temperature of nitrogen. Since BCS is a weak-coupling theory, it was found wanting to deal with such high-T_c SCs (HTSCs). It was also felt to be inadequate in accounting for the properties of a class of SCs, e.g., heavy-fermion SCs (HFSCs), for which therefore the term *exotic* or *unconventional* was coined.

We show in this monograph that a generalization of BCS equations (GBCSEs) enables one to address the superconducting features of non-elemental SCs in the manner elemental SCs are dealt with in the original theory. To elaborate, given Debye temperature of an elemental SC and its T_c, BCS theory enables one to predict the value of its gap Δ_0 at $T = 0$, or vice versa. Likewise, given the Debye temperature of a non-elemental

SC and any two parameters of the set $\{T_c, \Delta_{10}, \Delta_{20} > \Delta_{10}\}$, GBCSEs enable one to predict the remaining parameter. The desired generalization is achieved by adopting the "language" of Bethe–Salpeter equation (BSE). The importance of the choice of a language for description of physical phenomena cannot be over-emphasized. To state that dealing with the hydrogen atom by employing Cartesian coordinates would be *clumsy* is to state the obvious. A more pertinent example is provided by tensors which afford the most economical and elegant language for the general theory of relativity. To suggest that BSE provides a similar vehicle for superconductivity would seem to be bizarre, if not ridiculous, because the relativistic domain for which BSE was invented and the domain of superconductivity are as far apart as the north and the south poles.

To obtain results valid in the non-relativistic domain from a relativistic equation, however, is not really a problem since, to give a well-known example, one can obtain Schrodinger equation from Dirac equation by making appropriate approximations. To address superconductivity, we shall likewise resort to the non-relativistic approximation. This raises the legitimate question: Why then bother with the all the complexities of a 4-dimensional BSE? Salient advantages of adopting the BSE-based approach are noted below.

(a) It enables one at the outset — via the Matsubara technique — to temperature-generalize the pairing equation. The equation so obtained leads to the expression for the binding energy in the celebrated Cooper problem in the $T = 0$ limit; besides, with an appropriate change of limits, it leads also to

(b) The BCS equation for T_c of an elemental SC and to an alternative equation for $\Delta(T)$. More importantly,

(c) It leads to a manifest generalization of BCS equations which enable one to address HTSCs in the manner elemental SCs are dealt with in the original theory. This is achieved by employing for the kernel of the T-generalized BSE a superpropagator — in lieu of the usual one-phonon propagator employed for elemental SCs. As will be seen, we are then enabled to quantitatively address the observed high-T_cs and multiple gaps of HTSCs.

(d) It leads to new dynamics-based equations for the critical magnetic field and critical current density of an elemental and a non-elemental SC.

The presentation in this monograph has been inspired by two celebrated books, one authored by Sakurai[1] and the other by Dirac.[2] The former deals with an *enormous* task (major advances in the fundamentals of quantum physics from 1927 to 1967) and claims to do so "in a manner that cannot be made any simpler" — a claim that is amply justified. The latter book presents "the indispensible material in a direct and concise form" on general theory of relativity so as to enable students to pass through "the main obstacles" "with the minimum expenditure of time."

At the root of this monograph are quantum field theory (QFT) and its finite-temperature version — finite-temperature field theory (FTFT). Since the former *per se* is a formidable subject, and latter perhaps even more so, it seems imperative that we specify the level of preparation on the part of the reader to be able to easily follow the contents herein. The good news is that QFT is needed here to the extent of obtaining BSE Eq. (2.1) — no more! Since intuitive considerations that lead to Eq. (2.1) have been dealt with in Chapter 1 in a step-by-step manner, these can easily be followed by even those who have only a cursory familiarity with Feynman diagrams. Similarly, FTFT is needed here to the extent that it provides the recipe given in Eqs. (2.12) — no more. It will be seen that learning the use of these additional tools is a small price to pay for the rich dividend it yields: Unified treatment of various features of the superconductivity of both elemental and non-elemental SCs. The literature on QFT and FTFT is huge and, for the uninitiated, can be rather intimidating. In order not to overwhelm the reader for reasons already stated, references given herein are sparse. While due deference has been paid to classics in the field, referencing to other sources has been kept at a minimum and, as always, reflects the bias (or ignorance) of the author.

The author regards the rather unusual choice of topics covered in this monograph an important feature of it. While this reflects his own interests, he believes that it will be seen that there is a logical continuity to the topics that have been dealt with. He should also like to add that he has been privileged

---

[1] J.J. Sakurai, *Advanced Quantum Mechanics* (Addison Wesley; Boston, 1967).
[2] P.A.M. Dirac, *General Theory of Relativity* (John Wiley, NY, 1975).

to know serious workers who have devoted a lifetime to the field, but have been reluctant to go beyond certain confined areas of it. He hopes that this monograph will promote a widening of interest among those interested in this fascinating and challenging field.

The work reported here is a labour of love: Most of it was done post-retirement — without financial support from any agency. While he owes a debt of gratitude to many people for reasons acknowledged elsewhere, he would specially like to thank Lalit Pande, his friend and colleague for more than three decades, for inspiring him to *dig deep*, Manuel de Llano, for innumerable e-exchnges and for being an inspiration, and Arun Attri, who was a student when the author joined JNU and subsequently his colleague, for the gift of a Math software *after* which the author learnt to supplement his analytic work with numerical work, without which this monograph would not have been possible.

G.P. Malik

B-208 Sushant Lok I
Gurgaon 122009, Haryana, India
July 2015

More Frequently Used Abbreviations in the Text

BCS: Bardeen, Cooper, and Schrieffer
BEC: Bose–Einstein condensation
Bi-2212: $Bi_2Sr_2CaCu_2O_8$
BSE: Bethe–Salpeter equation
CP: Cooper pair
FTFT: Finite-temperature field theory
GBCSEs: Generalized BCS equations
HFSC: Heavy-fermion superconductor
HTSC: High-T_c superconductor
IA: Instantaneous approximation
LCO: La_2CuO_4
MDTs: Multiple Debye temperatures
MFA: Mean-field approximation
OPEM: One-phonon exchange mechanism
QFT: Quantum field theory
SC: Superconductor
Tl-2212: $Tl_2Ba_2CaCu_2O_8$
TPEM: Two-phonon exchange mechanism
YBCO: $YBa_2Cu_3O_8$

Symbols Used in More than One Chapter

β:	$1/k_B T$
$\Delta(T)$:	BCS energy gap at temperature T
Δ_0:	BCS energy gap at T = 0
Δ_{10}:	Smaller of the two gaps of an SC at T = 0
Δ_{20}:	Larger of the two gaps of an SC at T = 0
Δ_F:	Feynman propagator for a scalar field
E_F:	Fermi energy
k_B:	Boltzmann constant
$H_c(T)$:	Critical magnetic field at temperature T
H_0:	Critical magnetic field at $T = 0$
κ:	Thermal conductivity
κ_{es}:	Electronic part of thermal conductivity in the superconducting state
κ_{gs}:	Lattice part of thermal conductivity in the superconducting state
j_c:	Critical current density
j_0:	Critical current density at $T = 0$
λ_A:	Dimensionless BCS interaction parameter ([N(0)V]) of an elemental SC A
λ_m:	Dimensionless BCS interaction parameter for an elemental SC in an external magnetic field

λ_A^c: Dimensionless BCS interaction parameter of a non-elemental
            SC of which element A is a constituent

μ: Chemical potential

μ_0:   Chemical potential at $T = 0$

μ_1:   Chemical potential at $T = T_c$

$N(0)$:   Density of states at the Fermi surface for one spin, in units
            of $\text{eV}^{-1}\text{cm}^{-3}$

p_μ: Relative 4-momentum of two particles bound together

P_μ: 4-momemtum of the centre-of mass of two particles bound
            together

T: Temperature

T_c: Critical temperature

θ_D: Debye temperature

θ_D^A: Debye temperature of A ions in their free state

θ_D^{Ac}:   Debye temperature of A ions when they occur as constituents
            of a composite material

$-V$:   BCS parameter for net attraction between a pair of electrons
            because of attraction due to the ion-lattice and Coulomb
            repulsion, in units of eVcm^3

$|W(T)|$:   (half) the binding energy of a Cooper pair at temperature T

$|W_0|$:   (half) the binding energy of a Cooper pair at $T = 0$; equals Δ_0
            in the limit of infinesimal λ

$|W_{10}|$:   (half) the binding energy of a Cooper pair at $T = 0$ in the
            OPEM scenario; to be identified with Δ_{10}

$|W_{20}|$:   (half) the binding energy of a Cooper pair at $T = 0$ in the TPEM
            scenario; to be identified with Δ_{20}

ω_c: Debye cut off frequency $(k_B\theta_D/\hbar)$

Units: cgs and Natural

1. Values of some physical and numerical constants

$$c = 2.99792458 \times 10^{10} \text{ cm sec}^{-1},$$
$$1 \text{ eV} = 1.6021892 \times 10^{-12} \text{ gm cm}^2 \text{ sec}^{-1} \text{ (erg)},$$
$$\hbar = 6.582173 \times 10^{-16} \text{ eV sec},$$
$$\hbar c = 1.973285 \times 10^{-5} \text{ eV cm}-,$$
$$e^2/\hbar c = 1/137.03604,$$
$$k_{\text{Boltzmann}} = 8.61735 \times 10^{-5} \text{ K}^{-1} \text{ eV},$$
$$m_{\text{electron}} = 0.5110034 \text{ MeV/c}^2.$$

2. With a_i ($i = 1, 2, 3$) as given below

$$a_1 = 5.60958616 \times 10^{32}, \quad a_2 = 5.06772886 \times 10^4,$$
$$a_3 = 1.51926689 \times 10^{15},$$

the following identities can easily be verified by substituting for eV, c and $\hbar$ the numerical values given above:

$$1 \text{ gm} = a_1 \text{ eV}(\mathbf{c}^{-2})$$
$$1 \text{ cm} = a_2 \text{ eV}^{-1}(\hbar\mathbf{c})$$
$$1 \text{ sec} = a_3 \text{ eV}^{-1}(\hbar).$$

It is seen from these relations that all physical properties derivable from them, e.g., momentum, force, pressure, charge, etc., can be expressed in terms of eV, c, and $\hbar$. As an example, the cgs unit of magnetic field (gauss) can be written as:

$$1\,\text{gauss} = \text{gm}^{1/2}\text{cm}^{-1/2}\,\text{sec}^{-1} = (a_1^{1/2}a_2^{-1/2}a_3^{-1})\ \text{eV}^2(\hbar c)^{-3/2}.$$

3. It follows from paragraph 2 that if one were to adopt (eV, c, $\hbar$) as the basic units in lieu of (gm, cm, sec) *and* choose $c = \hbar = 1$, then every physical property can be expressed as some power of eV *only*. In the resulting *natural system of units* (*NSU*), the dimensions of any term in an equation can easily be checked without carrying along the factors of c and $\hbar$ in any calculation.

4. We have employed NSU in this monograph. However, since appropriate conversion factors have been incorporated into the equations, one actually needs to use the familiar BCS units for the input variables, which therefore lead to the output too in the same units. To elaborate, input/output for V (as in [N(0)V]) is in terms of eV cm^3, and gauss for the magnetic field H, and so on.

Contents

Chapter 1

The Bethe–Salpeter Equation (BSE)

1.1. Introduction

For reasons spelled out in the Preface, this monograph addresses various features of superconductivity by adopting the framework of the Bethe–Salpeter equation (BSE).[1-3] It would therefore seem imperative that an account be given of the background which led to the invention of BSE in the first place — prior to pointing out how the equation may be customized for superconductivity. This chapter is intended to serve this dual purpose.

It is well known that Schrödinger equation provides a fairly accurate description of the bound states of two spin-zero particles if mass of one is much greater than that of the other. Similarly, if the particles are spin-1/2 particles, one can use Dirac equation which was invented in 1920s to meet the need to make quantum theory manifestly consistent with the special theory of relativity, i.e., to treat the space and time variables on the same footing.

Experiments carried out with cosmic rays in late 1940s and high-energy accelerators built in early 1950s led to discovery of many new particles that were found to be composites of other particles moving with speeds approaching the velocity of light. There was therefore a need to find a *relativistic* equation for bound states of two particles with arbitrary masses. BSE was invented to address this need.

Since a particle with a definite momentum can be *anywhere* — as follows from the uncertainty principle, BSE was arrived at by following an intuitive approach to quantum field theory (which assigns an amplitude for the location of a particle at each space-time point) based on Feynman diagrams. The building blocks of this approach are Feynman propagators. BSE was therefore formulated in terms of two kinds of propagators: A two-particle propagator for the particles bound together and a propagator for the field that brought about the bound state. One could then draw hosts of Feynman diagram, each one of them representing a possible physical process. It was realized at the outset that the amplitude corresponding to any *finite* number of such diagrams would be inadequate because the bound particles interact arbitrarily often and stay together for infinitely long periods. This meant that the usual perturbation theory would not work. While, in principle, an exact equation can be derived in terms of "fully-clothed" or exact propagators and summing up diagrams that take into account every conceivable process, such an equation seems at present to be of only ornamental value because it cannot be solved.

A pragmatic approach to the relativistic bound-state problem is to consider a *subset* which, despite containing an infinite number of diagrams, leads nevertheless to a tractable problem. Such a subset consists of ladder diagrams built with free single-particle propagators. We deal below with BSE in the ladder approximation for the simplest case where all the particles are spinless.

1.2.   BSE for Two Spin-zero Particles Interacting via Another Spin-zero Particle in the Ladder Approximation

We consider the interaction Hamiltonian

$$H_{int} = g_a \Phi_a^+ \Phi_a \varphi + g_b \Phi_b^+ \Phi_b \varphi, \qquad (1.1)$$

where $\Phi_{a,b}$ and $\Phi_{a,b}^+$ are, respectively, the destruction and creation operators for a, b; φ denotes the c field, and g_a and g_b are the two interaction parameters. The amplitude for the initial two-particle state (a, b) to go over to the final state (a', b') in the ladder approximation is the sum of the amplitudes for

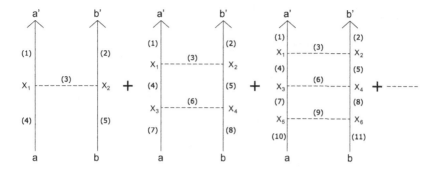

Fig. 1.1. Ladder diagrams.

the diagrams shown in Fig. 1.1. The amplitude for any diagram is obtained by multiplying together the amplitudes for each of its components. We now represent these components by symbols as follows: each open-ended component, e.g., 1, 2, 4, and 5 in the first diagram and 1, 2, 7, and 8 in the second diagram, by the symbol ⟨final state|Operator|initial state⟩; each closed-ended *cross* component, e.g., 3, 6, 9, etc., by the symbol D_F, and each closed-ended *non-cross* component, e.g., 4 and 5 in the second diagram and 3, 5, 7, and 8 in the third diagram, by the symbol Δ_{Fa} or Δ_{Fb}. D_F, Δ_{Fa}, and Δ_{Fb} are the propagators for c, a, and b, respectively. For the first diagram we then have

$(1) = \langle a'|\Phi_a^+(x_1)|0\rangle$, representing creation of a particle in state a' due to its creation operator acting on vacuum at x_1

$(2) = \langle b'|\Phi_b^+(x_2)|0\rangle$, representing creation of a particle in state b' due to its creation operator acting on vacuum at x_2

$(3) = D_F(x_1 - x_2)$, representing exchange of the c particle between a at x_1 and b at x_2 due to g_a and g_b

$(4) = \langle 0|\Phi_a(x_1)|a\rangle$, representing destruction into vacuum of a particle in state a due to its destruction operator at x_1

$(5) = \langle 0|\Phi_b(x_2)|b\rangle$, representing destruction into vacuum of a particle in state b due to its destruction operator at x_2

Multiplying the product of all the above factors by $(-ig_a)$ and $(-ig_b)$ because of the two vertices and integrating over the variables x_1 and x_2

of the internal line, the amplitude for this diagram is:

$$\langle a'b'|S^{(1)}|ab\rangle = \left[(-i)^2 g_a g_b \right.$$

$$\left. \times \int d^4x_1 d^4x_2 \langle a'|\Phi_a^+(x_1)|0\rangle \langle b'|\Phi_b^+(x_2)|0\rangle D_F(x_1 - x_2) \right]$$

$$\times \langle 0|\Phi_a(x_1)|a\rangle \langle 0|\Phi_b(x_2)|b\rangle.$$

Similarly, for the next diagram we have

$$\langle a'b'|S^{(2)}|ab\rangle = \left[(-i)^2 g_a g_b \right.$$

$$\left. \times \int d^4x_1 d^4x_2 \langle a'|\Phi_a^+(x_1)|0\rangle \langle b'|\Phi_b^+(x_2)|0\rangle D_F(x_1 - x_2) \right]$$

$$\times (-g_a g_b) \langle 0|\Phi_a(x_1)|a\rangle \langle 0|\Phi_b(x_2)|b\rangle$$

$$\times \int d^4x_3 d^4x_4 \Delta_{Fa}(x_1 - x_3) \Delta_{Fb}(x_2 - x_4) D_F(x_3 - x_4)$$

$$\times \langle 0|\Phi_a(x_3)|a\rangle \langle 0|\Phi_b(x_4)|b\rangle$$

The factors enclosed in the square brackets above occur not only in $\langle a'b'|S^{(1)}|ab\rangle$ and $\langle a'b'|S^{(2)}|ab\rangle$, but also in all the higher order terms. Therefore we have

$$\langle a'b'|S|ab\rangle = \langle a'b'|S^{(1)}|ab\rangle + \langle a'b'|S^{(2)}|ab\rangle + \cdots$$

$$= \left[(-i)^2 g_a g_b \right.$$

$$\left. \times \int d^4x_1 d^4x_2 \langle a'|\Phi_a^+(x_1)|0\rangle \langle b'|\Phi_b^+(x_2)|0\rangle D_F(x_1 - x_2) \right]$$

$$\times \left[\langle 0|\Phi_a(x_1)|a\rangle \langle 0|\Phi_b(x_2)|b\rangle (-g_a g_b) \right.$$

$$\times \int d^4x_3 d^4x_4 \Delta_{Fa}(x_1 - x_3) \Delta_{Fb}(x_2 - x_4) D_F(x_3 - x_4)$$

$$\left. \times \langle 0|\Phi_a(x_3)|a\rangle \langle 0|\Phi_b(x_4)|b\rangle + \cdots \right].$$

Since the expression within the second square brackets in the above expression is a function of x_1 and x_2, we may define

$$\chi(x_1 x_2 : ab) = \langle 0|\Phi_a(x_1)|a\rangle\langle 0|\Phi_b(x_2)|b\rangle + (-g_a g_b)$$

$$\times \int d^4 x_3 d^4 x_4 \Delta_{Fa}(x_1 - x_3)\Delta_{Fb}(x_2 - x_4)D_F(x_3 - x_4)$$

$$\times \langle 0|\Phi_a(x_3)|a\rangle\langle 0|\Phi_b(x_4)|b\rangle + \cdots .$$

$$(1.2)$$

It is seen that the above series in akin to a geometric series because each term in it, after the first, is obtained by multiplying the previous term by a double integral over the product $\Delta_{Fa}\Delta_{Fb}D_F$. Therefore we have

$$\chi(x_1 x_2 : ab) = \langle 0|\Phi_a(x_1)|a\rangle\langle 0|\Phi_b(x_2)|b\rangle + (-g_a g_b) \int d^4 x_3 d^4 x_4$$

$$\Delta_{Fa}(x_1 - x_3)\Delta_{Fb}(x_2 - x_4)D_F(x_3 - x_4)\chi(x_3 x_4 : ab).$$

$$(1.3)$$

Integral Eq. (1.3) is known as BSE.

1.3.   BSE for Scalar Particles in Momentum Space

For the bound states of the two particles, the inhomogeneous term in Eq. (1.3) must be dropped because it represents two non-interacting particles whence

$$\chi(x_1 x_2 : B) = (-g_a g_b) \int d^4 x_3 d^4 x_4 \Delta_{Fa}(x_1 - x_3)$$

$$\times \Delta_{Fb}(x_2 - x_4)D_F(x_3 - x_4)\chi(x_3 x_4 : B), \qquad (1.4)$$

where B denotes bound state. With applications to superconductivity in mind, we let $m_a = m_b = m$ and $g_a = g_b = g$. We now adopt the centre-of-mass and relative coordinates and go over to momentum space. This is done by introducing

$$X = \frac{1}{2}(x_1 + x_2), \quad x = x_1 - x_2, \quad P = p_1 + p_2, \quad p = \frac{1}{2}(p_1 - p_2),$$

$$(1.5)$$

defining

$$\chi(x_1 x_2 : B) = \exp(iPX) \int d^4 q \exp(iqx) \psi(q), \qquad (1.6)$$

and using the following momentum space representations of $\Delta_{Fa,Fb}$ and D_F:

$$\Delta_{Fa,Fb}(x) = \frac{-i}{(2\pi)^4} \int d^4 k \, G_{a,b}(k) \exp(ikx),$$

$$D_F(x) = \frac{-i}{(2\pi)^4} \int d^4 k \, D(k) \exp(ikx), \qquad (1.7)$$

where

$$G_{a,b}(k) = \frac{1}{k^2 + m_{a,b}^2}, \quad D(k) = \frac{1}{k^2 - i\varepsilon} \qquad (1.8)$$

and c is assumed to be massless.

Using Eqs. (1.6) and (1.7), we obtain

$$RHS \text{ of Eq. } (1.4) = (-g^2) \left[\frac{-i}{(2\pi)^4} \right]^3$$

$$\times \int d^4 x_3 d^4 x_4 d^4 k_1 d^4 k_2 d^4 k_3 d^4 p \, G_a(k_1) G_b(k_2) D(k_3) \psi(p)$$

$$\times \exp[ik_1(x_1 - x_3)] \exp[ik_2(x_2 - x_4)] \exp[ik_3(x_3 - x_4)]$$

$$\times \exp[ip(x_3 - x_4)] \exp\left[i\frac{P}{2}(x_3 + x_4)\right]$$

$$= (-g^2) \frac{i}{(2\pi)^4} \int d^4 k_1 d^4 k_2 d^4 k_3 d^4 p \, G_a(k_1) G_b(k_2) D(k_3) \psi(p)$$

$$\times \exp[i(k_1 x_1 + k_2 x_2)]$$

$$\times \left\{ \frac{1}{(2\pi)^4} \int d^4 x_3 \exp\left[ix_3\left(-k_1 + k_3 + p + \frac{P}{2}\right)\right] \right\}$$

$$\times \left\{ \frac{1}{(2\pi)^4} \int d^4 x_4 \exp\left[ix_4\left(-k_2 - k_3 - p + \frac{P}{2}\right)\right] \right\}$$

$$= (-g^2) \frac{i}{(2\pi)^4} \int d^4k_1 d^4k_2 d^4k_3 d^4p \, G_a(k_1) G_b(k_2) D(k_3) \psi(p)$$

$$\times \exp[i(k_1 x_1 + k_2 x_2)]$$

$$\times \delta^4 \left(-k_1 + k_3 + p + \frac{P}{2} \right) \delta^4 \left(-k_2 - k_3 - p + \frac{P}{2} \right)$$

$$= (-g^2) \frac{i}{(2\pi)^4} \int d^4k_3 d^4p \, G_a \left(k_3 + p + \frac{P}{2} \right)$$

$$G_b \left(-k_3 - p + \frac{P}{2} \right) D(k_3) \psi(p) \exp \left[i \left(k_3 + p + \frac{P}{2} \right) x_1 \right.$$

$$\left. + i \left(-k_3 - p + \frac{P}{2} \right) x_2 \right]$$

$$= \exp \left[\frac{iP}{2}(x_1 + x_2) \right] (-g^2) \frac{i}{(2\pi)^4}$$

$$\times \int d^4k_3 d^4p \, G_a \left(k_3 + p + \frac{P}{2} \right) G_b \left(-k_3 - p + \frac{P}{2} \right)$$

$$D(k_3) \psi(p) \exp[i(k_3 + p)(x_1 - x_2)]$$

$$= \exp \left[\frac{iP}{2}(x_1 + x_2) \right] (-g^2) \frac{i}{(2\pi)^4} \int d^4p \, d^4q$$

$$G_a \left(\frac{P}{2} + q \right) G_b \left(\frac{P}{2} - q \right) D(q - p) \psi(p) \exp[iq(x_1 - x_2)]$$

$$= \exp \left(\frac{iPX}{2} \right) (-g^2) \frac{i}{(2\pi)^4} \int d^4p \, d^4q \, G_a \left(\frac{P}{2} + q \right)$$

$$G_b \left(\frac{P}{2} - q \right) D(q - p) \psi(p) \exp(iqx). \tag{1.9}$$

Upon using Eqs. (1.6) and (1.9), we have Eq. (1.4) as

$$\int d^4q \exp(iqx) \left[\psi(q) + \frac{ig^2}{(2\pi)^4} \right.$$

$$\left. \times \int d^4p \, G_a \left(\frac{P}{2} + q \right) G_b \left(\frac{P}{2} - q \right) D(q - p) \psi(q) \right] = 0,$$

$$\tag{1.10}$$

or

$$\psi(q) = \frac{-ig^2}{(2\pi)^4} \int d^4 p \frac{1}{(P/2+q)^2 + m^2} \frac{1}{(P/2-q)^2 + m^2} I(q-p)\psi(p),$$

(1.11)

where $D(q-p) = 1/[(q-p)2 - i\varepsilon]$ has been replaced by $I(q-p)$, which is called the kernel of the equation and has different forms in different theories.

1.4. Remarks

1. Four-dimensional Eq. (1.11) is very difficult to solve even though it corresponds to an infinite number of a rather simple subset consisting of ladder diagrams. To this day, this equation remains the only BSE that has been solved in closed form. This solution was obtained by Wick.[4]

2. It can be seen from Eq. (1.3), which is equivalent to Eq. (1.11), that it corresponds to two time variables and therefore leads to conceptual difficulties associated with the physical interpretation of the BS wavefunction.

3. Equation (1.11) provides the springboard to deal with superconductivity. For this purpose we first need to obtain an equivalent equation when the particles bound together are fermions. Such an equation then requires customization to deal with superconductivity. We take up this task in the next chapter.

Notes and References

1. The original papers on BSE are:
   E.E. Salpeter and H.A. Bethe, *Phys. Rev.* **84**, 749 (1951);
   E.E. Salpeter, *Phys. Rev.* **87**, 328 (1952).
2. The treatment of BSE given here is based on Chapter 7 in:
   K. Nishijima, *Fields and Particles* (W.A. Benjamin, New York, 1969).
3. More formal and comprehensive treatments of BSE can be found in:
   M. Gell-Mann and F.E. Low, *Phys. Rev.* **84**, 350 (1951);
   D. Lurié, *Particles and Fields* (Interscience Publishers, New York, 1968);
   W. Greiner and J. Reinhardt, *Quantum Electrodynamics* (Springer-Verlag, Berlin, 1992).
4. G.C. Wick, *Phys. Rev.* **96**, 1124 (1954).
5. Comprehensive bibliography on BSE (till 1988) is given in the following two reviews:
   N. Nakanishi, *Prog. Theor. Phys.* **5**, 614 (1969).
   N. Nakanishi, *Prog. Theor. Phys. Suppl.* **95**, 10 (1988); bibliography on p. 78.

Chapter 2

Customization of the
Bethe–Salpeter Equation (BSE)
to Superconductivity

2.1. Introduction

It is universally agreed that superconductivity arises as a consequence of the formation of Cooper pairs, which are bound states of two electrons. Therefore in this chapter we first obtain a BSE for the bound states of two fermions. In order to customize this equation to deal with superconductivity, we introduce the important concept of instantaneous approximation (IA) without specifying the kernel of the equation. The equation so obtained is then temperature-generalized via the Matsubara technique.[1] Stage is thus set to apply, in the remaining chapters, the T-generalized BSE to both elemental and composite SCs by appropriately choosing its kernel in the mean-field approximation.

2.2. BSE for Two Spin-1/2 Particles Interacting via a Spin-zero Particle in the Ladder Approximation[2]

It is evident that to obtain the equation for spin-1/2 particles, we need to replace the boson propagators in Eq. (1.11) by fermion propagators.

9

On doing so, the resulting equation may be written as

$$\psi(p_\mu) = (-2\pi i)^{-1} \int d^4 q_\mu \frac{1}{\left[\gamma_\mu^{(1)} P_\mu/2 + \gamma_\mu^{(1)} q_\mu - m + i\varepsilon\right]}$$

$$\times \frac{1}{\left[\gamma_\mu^{(2)} P_\mu/2 - \gamma_\mu^{(2)} q_\mu - m + i\varepsilon\right]} I(q_\mu - p_\mu)\psi(q_\mu),$$

$$(2.1)$$

where P_μ is the total 4-momentum of the centre-of-mass of the particles, q_μ their relative 4-momentm, and $\gamma_\mu s$ are Dirac matrices for the two particles.

2.2.1.   *IA*

An essential step in the application of BSE to superconductivity is the adoption of IA which is concerned with using a reduced version of the kernel $I(q_\mu - p_\mu)$ in Eq. (2.1). Without IA, the kernel is a function of both the energy and 3-momentum parts. Adoption of IA means that the energy component of the kernel is ignored. Recalling that the kernel is the momentum-space transform of the propagator, which depends on both $(t_1 - t_2)$ and $(x_1 - x_2)$, this is equivalent to ignoring the time of travel of the particle that is exchanged between the particles it binds. Thus we set

$$I(q_\mu - p_\mu) = I(\mathbf{q} - \mathbf{p}). \tag{2.2}$$

Adoption of Eq. (2.2) not only leads to a tractable BSE suitable for superconductivity, but also rids the theory of the conceptual problem concerned with the interpretation of amplitudes for the bound state that depend on two time variables that was noted in the last chapter.

2.2.2.   *BSE in IA*

We now write Eq. (2.1) via a redefinition of ψ as

$$\left(\gamma_\mu^{(1)} P_\mu/2 + \gamma_\mu^{(1)} p_\mu - m + i\varepsilon\right) \left(\gamma_\mu^{(2)} P_\mu/2 - \gamma_\mu^{(2)} p_\mu - m + i\varepsilon\right) \chi(p_\mu)$$

$$= (-2\pi i)^{-1} \int d^4 q_\mu I(\mathbf{q} - \mathbf{p}) \chi(q_\mu) \tag{2.3}$$

and work in the rest frame of the two particles which means that $P_\mu = (E, 0)$; the metric employed is time-preferred: $a_\mu b_\mu = a_4 b_4 - \mathbf{a}.\mathbf{b}$. We now carry out spin-reduction of Eq. (2.3) by multiplying it with $\gamma_4^{(1)} \gamma_4^{(2)}$; then

$$\gamma_4^{(1)} \left(\gamma_\mu^{(1)} P_\mu / 2 + \gamma_\mu^{(1)} p_\mu - m + i\varepsilon \right) = \gamma_4^{(1)} \left(\gamma_4^{(1)} E/2 + \gamma_4^{(1)} p_4 \right.$$
$$\left. -\gamma^{(1)} \cdot \mathbf{p} - m + i\varepsilon \right)$$
$$= [E/2 + p_4 - H_1(\mathbf{p})],$$

where the following relations have been: $(\gamma_4^{(1)})^2 = 1$, $\boldsymbol{\gamma}^{(1)} = \beta^{(1)} \boldsymbol{\alpha}^{(1)}$, $\gamma_4^{(1)} = \beta^{(1)}$, $H_1(\mathbf{p}) = \mathbf{p} \cdot \boldsymbol{\alpha}^{(1)} + m\beta^{(1)}$; $H_1(\mathbf{p})$ being the Dirac Hamiltonian for the first particle in terms of the more familiar Dirac matrices β and α. $\gamma_4^{(2)}$ times the second factor on the LHS of Eq. (2.3) can be similarly dealt with, leading to

$$[E/2 + p_4 - H_1(\mathbf{p})][E/2 - p_4 - H_2(\mathbf{p})]\chi(p_\mu)$$
$$= (-2\pi i)^{-1} \int d^4 q_\mu I(\mathbf{q} - \mathbf{p})\chi(q_\mu), \qquad (2.4)$$

where $\gamma_4^{(1)} \gamma_4^{(2)}$ on the RHS has been absorbed in the kernel $I(\mathbf{q}-\mathbf{p})$.

Equation (2.4) is a complicated matrix equation in which $H_1(\mathbf{p})$ and $H_2(\mathbf{p})$ are given by 4×4 matrices. However, since we are interested in its applications in the extreme non-relativistic limit, we can simplify it by the use of Casimir projection operators defined for particle 1 as:

$$\Lambda_+^{(1)}(\mathbf{p}) = \frac{\Sigma_1(\mathbf{p}) + H_1(\mathbf{p})}{2\Sigma_1(\mathbf{p})}, \qquad \Lambda_-^{(1)}(\mathbf{p}) = \frac{\Sigma_1(\mathbf{p}) - H_1(\mathbf{p})}{2\Sigma_1(\mathbf{p})},$$

$$\Sigma_1(\mathbf{p}) = +\sqrt{\mathbf{p}^2 + m^2},$$

with similar definitions for particle 2. Further, we may regard $\chi(p_\mu)$ as a product of the individual wave functions of the two particles:

$$\chi(p_\mu) = \chi^{(1)}(p_\mu)\chi^{(2)}(p_\mu) = \left[\chi_+^{(1)}(p_\mu) + \chi_-^{(1)}(p_\mu) \right] \left[\chi_+^{(2)}(p_\mu) + \chi_-^{(2)}(p_\mu) \right].$$

We then have

$$\Lambda_+^{(1)}\chi(p_\mu) = \chi_+^{(1)}(p_\mu),$$

and similarly for the second particle.

Subjecting Eq. (2.4) to the operation of $\Lambda_+^{(1)}\Lambda_+^{(2)}$ we have

$$\Lambda_+^{(1)}\Lambda_+^{(2)}LHS \text{ of Eq. (2.4)} = \Lambda_+^{(1)}[E/2 + p_4 - H_1(\mathbf{p})]\chi^{(1)}(p_\mu)$$

$$\times \Lambda_+^{(2)}[E/2 - p_4 - H_2(\mathbf{p})]\chi^{(2)}(p_\mu)$$

$$= [E/2 + p_4 - H_1(\mathbf{p})]\Lambda_+^{(1)}\chi^{(1)}(p_\mu)$$

$$\times [E/2 - p_4 - H_2(\mathbf{p})]\Lambda_+^{(2)}\chi^{(2)}(p_\mu)$$

$$= [E/2 + p_4 - H_1(\mathbf{p})]\chi_+^{(1)}(p_\mu)$$

$$\times [E/2 - p_4 - H_2(\mathbf{p})]\chi_+^{(2)}(p_\mu)$$

$$= [E/2 + p_4 - \Sigma_1(\mathbf{p})]$$

$$\times [E/2 - p_4 - \Sigma_2(\mathbf{p})]\chi_{++}(p_\mu), \quad (2.5\ a)$$

where

$$\chi_{++}(p_\mu) = \chi_+^{(1)}(p_\mu)\chi_+^{(2)}(p_\mu) = \Lambda_+^{(1)}(\mathbf{p})\Lambda_+^{(2)}(\mathbf{p})\chi(p_\mu). \quad (2.6\ a)$$

Equations similar to Eqs. (2.5 a) and (2.6 a) obtained when Eq. (2.4) is subjected to the operation of $\Lambda_+^{(1)}\Lambda_-^{(2)}$, $\Lambda_-^{(1)}\Lambda_+^{(2)}$, and $\Lambda_-^{(1)}\Lambda_-^{(2)}$, respectively, are

$$\Lambda_+^{(1)}\Lambda_-^{(2)} \text{ LHS of Eq. (2.4)}$$

$$= [E/2 + p_4 - \Sigma_1(\mathbf{p})][E/2 - p_4 + \Sigma_2(\mathbf{p})]\chi_{+-}(p_\mu)$$

$$\Lambda_-^{(1)}\Lambda_+^{(2)} \text{ LHS of Eq. (2.4)}$$

$$= [E/2 + p_4 + \Sigma_1(\mathbf{p})][E/2 - p_4 - \Sigma_2(\mathbf{p})]\chi_{-+}(p_\mu)$$

$$\Lambda_-^{(1)}\Lambda_-^{(2)} \text{ LHS of Eq. (2.4)}$$

$$= [E/2 + p_4 + \Sigma_1(\mathbf{p})][E/2 - p_4 + \Sigma_2(\mathbf{p})]\chi_{--}(p_\mu)$$

$$(2.5\ b, c, d)$$

and

$$\chi_{+-}(p_\mu) = \Lambda_+^{(1)}(\mathbf{p})\Lambda_-^{(2)}(\mathbf{p})\chi(p_\mu)$$

$$\chi_{-+}(p_\mu) = \Lambda_-^{(1)}(\mathbf{p})\Lambda_+^{(2)}(\mathbf{p})\chi(p_\mu) \quad\quad (2.6\ b, c, d)$$

$$\chi_{--}(p_\mu) = \Lambda_-^{(1)}(\mathbf{p})\Lambda_-^{(2)}(\mathbf{p})\chi(p_\mu).$$

Multiplying the RHS of Eq. (2.4) with $\Lambda_+^{(1)}(\mathbf{p})\Lambda_+^{(2)}(\mathbf{p})$ and using Eqs. (2.5) and (2.6) we obtain

$$
[E/2 + p_4 - \Sigma_1(\mathbf{p})][E/2 - p_4 - \Sigma_2(\mathbf{p})]\chi_{++}(p_\mu)
$$
$$
= (-2\pi i)^{-1}\Lambda_+^{(1)}(\mathbf{p})\Lambda_+^{(2)}(\mathbf{p})
$$
$$
\times \int d^4 q_\mu I(\mathbf{q} - \mathbf{p})\left[\Lambda_+^{(1)}(\mathbf{q})\Lambda_+^{(2)}(\mathbf{q})\right]^{-1}\chi_{++}(q_\mu),
$$

or

$$
[\Lambda_+^{(1)}(\mathbf{p})\Lambda_+^{(2)}(\mathbf{p})]^{-1}[E/2 + p_4 - \Sigma_1(\mathbf{p})][E/2 - p_4 - \Sigma_2(\mathbf{p})]\chi_{++}(p_\mu)
$$
$$
= (-2\pi i)^{-1}\int d^4 q_\mu I(\mathbf{q} - \mathbf{p})[\Lambda_+^{(1)}(\mathbf{q})\Lambda_+^{(2)}(\mathbf{q})]^{-1}\chi_{++}(q_\mu). \quad (2.7)
$$

Note that RHS of this equation is a function of $\mathbf{p}$ alone; consistency demands that so must be the LHS. Therefore we define

$$
[\Lambda_+^{(1)}(\mathbf{p})\Lambda_+^{(2)}(\mathbf{p})]^{-1}[E/2 + p_4 - \Sigma_1(\mathbf{p})]
$$
$$
\times [E/2 - p_4 - \Sigma_2(\mathbf{p})]\chi_{++}(p_\mu) = S_{++}(\mathbf{p}),
$$

so as to obtain Eq. (2.7) as

$$
S_{++}(\mathbf{p}) = (-2\pi i)^{-1}\int d^3\mathbf{q} I(\mathbf{q} - \mathbf{p})S_{++}(\mathbf{q}) \cdot J_{++}(\mathbf{q}), \quad (2.8\,\mathrm{a})
$$

where

$$
J_{++}(\mathbf{q}) = \int \frac{dq_4}{[E/2 + q_4 - \Sigma(\mathbf{q})][E/2 - q_4 - \Sigma(\mathbf{q})]}, \quad (2.9\,\mathrm{a})
$$

where we have put $\Sigma_1(\mathbf{q}) = \Sigma_2(\mathbf{q}) = (\mathbf{q}^2 + m^2)^{1/2} \equiv \Sigma(\mathbf{q})$ since we are considering the bound state of equal-mass particles.

Equations similar to Eqs. (2.8 a) and (2.9 a) obtained when Eq. (2.4) is subjected to the operation of $\Lambda_+^{(1)}\Lambda_-^{(2)}$, $\Lambda_-^{(1)}\Lambda_+^{(2)}$, and $\Lambda_-^{(1)}\Lambda_-^{(2)}$, respectively, are

$$S_{+-}(\mathbf{p}) = (-2\pi i)^{-1} \int d^3\mathbf{q} I(\mathbf{q} - \mathbf{p}) S_{+-}(\mathbf{q}) \cdot J_{+-}(\mathbf{q})$$

$$S_{-+}(\mathbf{p}) = (-2\pi i)^{-1} \int d^3\mathbf{q} I(\mathbf{q} - \mathbf{p}) S_{-+}(\mathbf{q}) \cdot J_{-+}(\mathbf{q}) \qquad (2.8\ b,\ c,\ d)$$

$$S_{--}(\mathbf{p}) = (-2\pi i)^{-1} \int d^3\mathbf{q} I(\mathbf{q} - \mathbf{p}) S_{--}(\mathbf{q}) \cdot J_{--}(\mathbf{q}),$$

where

$$J_{+-}(\mathbf{q}) = \int \frac{dq_4}{[E/2 + q_4 - \Sigma(\mathbf{q})][E/2 - q_4 + \Sigma(\mathbf{q})]}$$

$$J_{-+}(\mathbf{q}) = \int \frac{dq_4}{[E/2 + q_4 + \Sigma(\mathbf{q})][E/2 - q_4 - \Sigma(\mathbf{q})]} \qquad (2.9\ b,\ c,\ d)$$

$$J_{--}(\mathbf{q}) = \int \frac{dq_4}{[E/2 + q_4 + \Sigma(\mathbf{q})][E/2 - q_4 + \Sigma(\mathbf{q})]}.$$

We deal below with the considerations that lead via the application of Matsubara technique to Eqs. (2.9 a–d) to the temperature-generalization of Eqs. (2.8 a–d)).

2.3. Temperature-generalization of $T = 0$ BSE via Matsubara Technique

In the decade in which BCS theory and BSE were created, 1950s, took place another landmark theoretical development: Creation of finite-temperature field theory by Matsubara. The following brief account of this theory is intended to serve the limited purpose of making palatable our treatment of Eqs. (2.9 a–d) in order to introduce temperature into the theory. This necessitates consideration of statistical operators of the type

$$O = \exp[-\beta(H - \mu N)], \ (\beta = 1/k_B T), \qquad (2.10)$$

where the Hamiltonian H and the number operator N are taken in the second-quantization and Schrödinger representations, k_B is the Boltzmann constant and μ the chemical potential. With $H = H_0 + H'$, where H_0 is the free Hamiltonian and H' the perturbation Hamiltonian, differentiating Eq. (2.10) with respect to β we obtain

$$-\frac{dO}{d\beta} = (H_0 - \mu N + H')O. \qquad (2.11)$$

The crucial observation about Eq. (2.11) is that it is analogous to Schrödinger equation if *it* in it is replaced by β. With this observation as the starting point, Matsubara constructed a special interaction representation in terms of temperature. The usual S matrix is then replaced by $S(\beta)$, which describes evolution of the system in terms of β (or temperature). The Feynman rules of $T = 0$ theory can then be taken over to the $T \neq 0$ theory with *it* replaced by β. The basic difference between the two theories is that the limits of integration over β are finite. This leads one to continue β periodically along the entire β axis, enabling temperature-dependent free–particle Green functions to be defined as Fourier series in β. The upshot of these considerations is that the amplitude for any $T = 0$ Feynman diagram (such as those that lead to BSEs in this and the previous chapter), and hence an integral equation representing the sum of such diagrams, can be T-generalized via the following substitutions[3]:

$$
\begin{aligned}
q_4 &= (2n + 1)\pi/ - i\beta && \text{for fermions} \\
&= 2n\pi/ - i\beta && \text{for bosons} \\
\int dq_4 &= \frac{2\pi}{-i\beta} \sum_{n=-\infty}^{\infty}.
\end{aligned}
\qquad (2.12)
$$

2.4. Temperature-generalization of (2.8 a–d) via (2.9 a–d)

In order to temperature-generalize either of (2.8 a–d) via the corresponding equation in (2.9 a–d), consider the equation

$$J(\mathbf{q}) = \int \frac{dq_4}{(q_4 + A)(-q_4 + B)} = \frac{1}{(A + B)} \left[\int \frac{dq_4}{(A + q_4)} - \int \frac{dq_4}{(B - q_4)} \right].$$

Apply to this equation now the prescription for fermions given in Eq. (2.12):

$$J(\mathbf{q})|_\beta = \frac{1}{(A + B)} \left[\sum_n \frac{1}{n - i\beta A/2\pi + 1/2} - \sum_n \frac{1}{n + i\beta B/2\pi + 1/2} \right].$$

The summations in this equation may be carried out by using

$$\sum_{n=-\infty}^{\infty} f(n) = (-\pi)[\text{sum of residues of } f(z)\cot(\pi z) \text{ at the poles of } f(z)],$$

whence

$$J(q)|_\beta = \frac{-\pi}{(A+B)}[-\cot(\pi/2 - i\beta A/2) + \cot(\pi/2 + i\beta B/2)].$$

Upon using

$$\coth(x \pm iy) = \frac{\sinh(2x) \mp i\sin(2y)}{\cosh(2x) - \cos(2y)},$$

we have

$$J(\mathbf{q})|_\beta = \frac{-\pi}{(A+B)}\left[-\frac{i\sinh(\beta A)}{\cosh(\beta A) + 1} - \frac{i\sinh(\beta B)}{\cosh(\beta B) + 1}\right]$$

$$= \frac{i\pi}{(A+B)}[\tanh(\beta A/2) + \tanh(\beta B/2)] \tag{2.13}$$

We can now obtain expressions for $J_{++}(\mathbf{q})|_\beta, \ldots, J_{--}(\mathbf{q})|_\beta$ from Eq. (2.13) by choosing A and B appropriately. For the first of these, $A = B = E/2 - \Sigma(\mathbf{p})$, whence we obtain Eq. (2.8 a–d) as

$$S_{++}(\mathbf{p}) = (-1)\int_L^U d^3\mathbf{q}\,\frac{\tanh[(\beta/2)\{E/2 - \Sigma(\mathbf{q})\}]}{E - 2\Sigma(\mathbf{q})}I(\mathbf{q} - \mathbf{p})S_{++}(\mathbf{q}).$$

$$(2.14\,\text{a})$$

For $J_{+-}(\mathbf{q})|_\beta$, $A = E/2 - \Sigma(\mathbf{p})$ and $B = E/2 + \Sigma(\mathbf{p})$; therefore Eq. (2.8 b) is obtained as

$$S_{+-}(\mathbf{p}) = \left(-\frac{1}{2}\right)\int_L^U d^3\mathbf{q}$$

$$\times \frac{\tanh[(\beta/2)\{E/2 - \Sigma(\mathbf{q})\}] + \tanh[(\beta/2)\{E/2 + \Sigma(\mathbf{q})\}]}{E}$$

$$\times I(\mathbf{q} - \mathbf{p})S_{++}(\mathbf{q}). \tag{2.14 b}$$

Similarly we obtain

$$S_{-+}(\mathbf{p}) = \left(-\frac{1}{2}\right) \int_{L}^{U} d^3\mathbf{q}$$

$$\times \frac{\tanh[(\beta/2)\{E/2 + \Sigma(\mathbf{q})\}] + \tanh[(\beta/2)\{E/2 - \Sigma(\mathbf{q})\}]}{E}$$

$$\times I(\mathbf{q} - \mathbf{p})S_{-+}(\mathbf{q})$$

$$S_{--}(\mathbf{p}) = (-1) \int_{L}^{U} d^3\mathbf{q} \frac{\tanh[(\beta/2)\{E/2 + \Sigma(\mathbf{q})\}]}{E + 2\Sigma(\mathbf{q})} I(\mathbf{q} - \mathbf{p})S_{--}(\mathbf{q}).$$

$$(2.14 \text{ c, d})$$

The kernel $I(\mathbf{q}-\mathbf{p})$ in the above equations has not yet been specified. Its choice is governed by the problem one needs to tackle, which then also determines the limits (L, U). It will be shown in the next chapter that choosing the kernel in the above equations to be in accord with (i) the Cooper model interaction leads to the celebrated expression for the binding energy in the Cooper problem, (ii) the BCS model interaction leads to an equation identical with the BCS equation for T_c and an equation equivalent to the BCS gap equation for elemental SCs. The implications of the remaining equations will also be taken up in the next chapter. It will be shown in subsequent chapters that just *one* of the equations obtained in this chapter, Eq. (2.14 a), enables one to address several features of the superconductivity of both elemental and high-T_c SCs.

2.5. Remarks

Temperature-generalization of Eq. (2.1) leading to Eqs. (2.14 a–d) is a huge step: It converts the original 2-particle problem defined in vacuum to a many-body problem because temperature is a statistical concept. It will be shown in the next chapter that when $T = 0$, Eq. (2.14 a) leads to the familiar result of the 2-electron Cooper problem if electrons below the Fermi surface are excluded from taking part in the dynamics of pairing. If not, then even at $T = 0$ Eq. (2.14 a) defines a many-body problem, leading to an expression for $|W_0|$ which is equivalent to Δ_0 in BCS theory.

Notes and References

1. At the root of finite-temperature field theory (FTFT) is:
T. Matsubara, *Prog. Theor. Phys.* **14**, 351 (1955).
Some of the texts dealing with systematic development of FTFT are:
D.A. Kirzhnits, *Field Theoretical Methods in Many-Body Systems* (Pergamon, New York, 1967);
J.I. Kaputsa and C. Gale, *Finite-Temperature Field Theory Principles and Applications* (Cambridge University Press, Cambridge, 2006);
A. Das, *Finite Temperature Field Theory* (World Scientific, Singapore, 1997).
Classic books with emphasis on applications of FTFT to many-body physics are:
J.R. Schrieffer, *Theory of Superconductivity* (W.A. Benjamin, Reading, Massachusetts, 1964);
A.A. Abrikosov, L.P. Gorkov and I.E. Dzyaloshinski, *Methods of Quantum Field Theory in Statistical Physics* (Dover, New York, 1963);
A. L. Fetter and J.D. Walecka, *Quantum Theory of Many-Particle Systems* (McGraw Hill, New York, 1971).
2. The treatment in this chapter is based on
E.E. Salpeter, *Phys. Rev.* **87**, 328 (1952).
3. See, for example,
A.D. Linde, *Reps. Prog. Phys.* **42**, 389 (1979);
L. Dolan and R. Jackiw, *Phys. Rev. D* **9**, 3320 (1979);
S. Weinberg, *Phys. Rev. D* **9**, 3357 (1979).
4. Some of the results of this chapter were first obtained in:
G.P. Malik and U. Malik, *Physica B* **336**, 349 (2003);
G.P. Malik, *Physica C* **468**, 949 (2008);
G.P. Malik, *Int. J. Mod. Phys. B* **24**, 1159 (2010).

Chapter 3

Re-derivation of Some Well-Known Results of BCS Theory via BSE-Based Approach

3.1. Introduction

The key to the mystery of superconductivity was provided by solution of the 2-electron Cooper problem.[1] Creation of the successful BCS theory,[2] which deals with systems with electrons of the order of 10^{23}, from the considerations of such a problem was a huge step. In order to juxtapose the conventional approach to superconductivity with the BSE-based approach, we re-obtain in this chapter the major results of the Cooper problem and BCS theory via the latter approach. Before doing so, however, we also include here an account of the conventional approach — not as a substitute for the excellent accounts of it already available, but in order to put in focus the differences in the conceptual frameworks of the two approaches.

3.2. Cooper Problem

This problem was concerned with calculation of the change in the ground state energy of a free electron gas at $T = 0$ when an attractive interaction, *however small*, is postulated between a pair of electrons introduced at the top of the Fermi sea. It was found that this hypothetical situation leads to the formation of a bound state of the electrons with opposite spins and zero total momentum. This bound state is called a Cooper pair (CP); the ion-lattice plays an essential role in its formation since without it the electrons would repel each other. The following picture has often been given for origin of the attractive interaction: Passage of an electron through the crystal lattice leaves behind a deformation trail which can be regarded as an area of enhanced positive charges; if a second electron passes through the lattice while it is recovering from the effect of the passage of the first, it will feel an attractive force. Therefore, overall, the electron–lattice–electron interaction can be a net weak attractive interaction. In the language of quantum field theory, it is said that the two electrons exchange a phonon because of the lattice.

The quantitative features of the Cooper problem may be summarized as:

(i) Each electron is represented by a plane wave and the wave function of the pair is expressed as a sum over products of these:

$$\psi(\mathbf{r}_1 - \mathbf{r}_2) = \sum_k g(\mathbf{k}) \exp[i\mathbf{k}.(\mathbf{r}_1 - \mathbf{r}_2)], \quad g(\mathbf{k}) = 0 \quad \text{for } \mathbf{k} < \mathbf{k}_F.$$

$$(3.1)$$

In this equation only those states have been considered for which the centre of mass is at rest: If $\mathbf{k}$ be the wave vector of one electron in the pair, then that of the other is $-\mathbf{k}$; $g(\mathbf{k})$ is the probability amplitude for such states; $g(\mathbf{k}) = 0$ for $\mathbf{k} < \mathbf{k}_F$ follows from Pauli's exclusion principle because states below $\mathbf{k}_F$ are already occupied.

(ii) Without the postulated attractive interaction, total energy of the pair at the top of the Fermi surface would be

$$E = 2E_F + W, \quad (W > 0), \tag{3.2}$$

where $E_F = \hbar^2 k_F^2 / 2m$ is the Fermi energy up to which all levels are occupied, W the additional energy of the two electrons, and k_F and m are, respectively, the Fermi wave vector and the mass of an electron.

(iii) The postulated form of the attractive interaction is proportional to

$$-V(V > 0) \quad \text{for } E_F \leq p^2/2m \leq E_F + k_B\theta_D, \quad \text{zero otherwise;}$$

$$(3.3)$$

where $p = \hbar k$, k being the wave vector, and θ_D, the Debye temperature, is given by $\hbar\omega/k_B$, ω being the Debye frequency of vibrations of the ions.

(iv) Solution of the Schrödinger equation with the wave function given by Eq. (3.1) and the above interaction leads to

$$W = -2k_B\theta_D \exp(-2/[N(0)V]), \quad (3.4)$$

where it is assumed that $[N(0)V] \ll 1$ and $N(0)$ is the density of states per unit energy at the Fermi surface, given by $(2m)^{3/2} E_F^{1/2}/4\pi^2\hbar^3$.

It is therefore seen that interaction of the form of Eq. (3.3) between two electrons at the top of the Fermi sea at $T = 0$ causes a lowering of their total energy because of the negative sign of W. This is said to cause an instability in the initial state of the system, suggesting that more and more electrons should end up as CPs.

3.3. From Cooper Problem to Major Results of BCS Theory at $T = 0$: A Qualitative Account

3.3.1. *The reduced BCS wave function*

Since formation of a CP in the 2-electrom Cooper problem causes a reduction in the energy of the Fermi sea, one would expect a similar lowering of the ground state energy of a many-electron system via pairing. However, the total reduction cannot be obtained by simply multiplying the reduction due to a single pair by one-half the number of electrons in the system because the effect of each CP depends on those already present. The ground state of an SC is achieved through a complicated interaction involving electrons of the order of 10^{23} per cm^3. This is the formidable problem tackled by BCS theory.

To deal realistically with an SC, one needs the wave function and Hamiltonian for a *huge* number of particles. It was immediately realized that a simple generalization of Eq. (2.1) would not work for the reasons that (i) it would fix the number of pairs which ought to be variable and, even if it were not so, (ii) it would lead to an intractable problem. BCS theory tacked the first of these problems by introducing the concept of a reduced wave function based on the formalism of second quantization and the second problem by introducing a reduced Hamiltonian for the many-electron system.

The usual framework of second quantization employs creation and destruction operators which, acting on any state, increase or decrease the number of particles by one. The reduced BCS wave function, however, was written in terms of two-particle creation and destruction operators:

$$|\psi_{BCS}\rangle \approx \exp\left[\sum g_k b_k^+\right]|0\rangle$$

where g_ks are parameters to be adjusted so as to minimize the total energy, b_k^+ is the creation operator for a pair the constituents of which have momenta $\mathbf{k}$ and $-\mathbf{k}$ and opposite spins, and $|0\rangle$ is the state with completely filled Fermi sea, i.e., the ground state of a non-interacting electron gas with no electron or hole excitations. Note that the above wave function is a superposition of pair states containing $0, 2, 4, \ldots N$ electrons, i.e., it does not define a state with a fixed number of particles. Invoking the commutation relation satisfied by the operators $b_{\mathbf{k}}^+$ and $b_{\mathbf{k}'}^+$, the normalized form of the above wave function is found to be

$$|\psi_{BCS}\rangle_n = \prod(u_{\mathbf{k}} + v_{\mathbf{k}} b_{\mathbf{k}}^+)|0\rangle, \quad u_{\mathbf{k}}^2 + v_{\mathbf{k}}^2 = 1, \tag{3.5}$$

where $u_{\mathbf{k}} = 1/(1+g_{\mathbf{k}}^2)^{1/2}$ and $v_{\mathbf{k}} = g_k/(1+g_{\mathbf{k}}^2)^{1/2}$, the physical significance of $u_{\mathbf{k}}$ being that it is the amplitude for the paired state of two electrons with opposite momenta and spins to be unoccupied, whereas $v_{\mathbf{k}}$ is the amplitude for this state to be occupied.

3.3.2.   The reduced BCS Hamiltonian

The Hamiltonian of an SC may generally be assumed to have a term corresponding to kinetic energy of each of its electrons, a similar term for the phonons, a term due to Coulomb repulsion of electrons, and a term due to the electro–phonon interactions. This leads to a hopeless situation given the number of constituents of the system. There was a need therefore once more to *reduce* the task at hand, while retaining the essence of the physics involved. It was argued in BCS theory that in order to understand superconductivity, it should suffice to consider the piece of the general Hamiltonian which is responsible for the *difference* in the normal and superconducting phases. Such a simplification is suggested by the fact that many properties of the system remain unaltered in the transition from the normal to the superconducting phase. The reduced BCS Hamiltonian is:

$$H_{BCS} = 2 \sum \xi_k b_k^+ b_k - \sum V_{k,k'} b_k^+ b_k, \qquad (3.6)$$

where $\xi_k = \hbar^2 k^2 / 2m - \mu$; $\mu = E_F$ is included in this term which multiplies the number operator $b_k^+ b_k$ in order to regulate the *mean* number of particles. The second piece of the Hamiltonian has been written by assuming that the phonon degrees of freedom can be cast in the form of an effective electron-phonon interaction containing only two CPs.

3.3.3. *Determination of u_k and v_k and the equation for $T = 0$ gap*

These amplitudes are determined by first calculating the expectation value of H_{BCS} given by Eq. (3.6) in the state given by Eq. (3.5). This gives

$$\langle \psi_{BCS} | H_{BCS} | \psi_{BCS} \rangle = 2 \sum_k \xi_k v_k^2 + \sum_{k,k'} V_{kk'} u_k v_{k'}^+ u_{k'}^+ v_k.$$

It is now demanded that the first-order variation of this expectation value should vanish. This calculation is simplified by assuming u_k and v_k to be real and by writing $u_k = \sin \theta_k$ and $v_k = \cos \theta_k$ (which is suggested by the second of the two equations in Eq. 3.5) whence

$$\langle \psi_{BCS} | H_{BCS} | \psi_{BCS} \rangle = 2 \sum_k \xi_k \cos^2 \theta_k + \frac{1}{4} \sum_{k,k'} V_{kk'} \sin(2\theta_k) \sin(2\theta_{k'}).$$

$$(3.7)$$

Equating the derivative of this equation with respect to θ_k to zero, one obtains

$$\tan(2\theta_k) = \frac{\sum_p V_{pk} \sin(2\theta_p)}{2\xi_k}. \qquad (3.8)$$

At this stage the following definitions are introduced:

$$\Delta_k = -\frac{1}{2} \sum_i V_{ki} \sin(2\theta_i), \; E_k = \sqrt{\Delta_k^2 + \xi_k^2}, \qquad (3.9)$$

whence we have

$$\tan(2\theta_k) = -\Delta_k / \xi_k, \; 2u_k v_k = \sin(2\theta_k) = \Delta_k / E_k,$$
$$v_k^2 - u_k^2 = \cos(2\theta_k) = -\xi_k / E_k. \qquad (3.10)$$

Hence we obtain

$$u_{\mathbf{k}}^2 = \frac{1}{2}\left(1 + \frac{\xi_{\mathbf{k}}}{E_{\mathbf{k}}}\right), \quad v_{\mathbf{k}}^2 = \frac{1}{2}\left(1 - \frac{\xi_{\mathbf{k}}}{E_{\mathbf{k}}}\right)$$

and Eq. (3.7) as

$$\Delta_{\mathbf{k}} = -\frac{1}{2}\sum_i V_{ki}\frac{\Delta_i}{E_i}. \tag{3.11}$$

This equation is the BCS equation for the gap at $T = 0$. Deferring justification as to why it is so called to the next section, we now proceed to solve it. To this end, it is instructive to first plot the BCS occupation probability function, $P(\xi) = v_{\mathbf{k}}^2(\xi)$, given above. This plot differs from the plot corresponding to the situation when there is no pairing only in a very narrow region. This leads one to make the reasonable assumption that the interaction responsible for pairing may be treated as having a constant value in this narrow region which is taken to be $E_F \pm k_B\theta_D$:

$$V_{ki} = -V \quad \text{for } E_F - k_B\theta_D \leq p_k^2/2m, \ p_i^2/2m \leq E_F + k_B\theta_D (V > 0)$$
$$= 0, \quad \text{otherwise.}$$

$$\tag{3.12}$$

Note that the model BCS interaction above is the one employed earlier in the Cooper problem, except that the lower limit up to which it is operative is $E_F - k_B\theta_D$ rather than E_F. Replacing the summation over i in Eq. (3.11) by integration and using Eq. (3.12) we obtain

$$1 = \frac{V}{2(8\pi^3)}\int_{E_F - k_B\theta_D}^{E_F + k_B\theta_D} d^3p \frac{1}{\sqrt{(p^2/2m - E_F)^2 + \Delta^2}}, \tag{3.13}$$

since Δ is now a constant and $E_k = \sqrt{(\hbar^2 k^2/2m - E_F)^2 + \Delta^2} = \sqrt{(p^2/2m - E_F)^2 + \Delta^2}$.

With $\xi = p^2/2m - E_F$,

$$\frac{d^3p}{8\pi^3} = \frac{p^2dp}{2\pi^2} = \frac{(2m)^{3/2}(E_F + \xi)^{1/2}d\xi}{4\pi^2} = \frac{(2m)^{3/2}E_F^{1/2}d\xi}{4\pi^2} \equiv N(0)d\xi,$$

we have Eq. (3.13) as

$$1 = [N(0)V] \int_0^{k_B\theta_D} d\xi \frac{1}{(\xi^2 + \Delta^2)^{1/2}}, \qquad (3.13a)$$

since $\frac{\xi_{\max}}{E_F} \ll 1$. $N(0)$ above is the density of states per unit energy at the Fermi surface as in the Cooper problem. From Eq. (3.13) we now obtain

$$1 = [N(0)V]\text{arcsinh}\left(\frac{k_B\theta_D}{\Delta}\right),$$

which, in the weak coupling limit, $[N(0)V] \ll 1$, reduces to

$$\Delta = 2k_B\theta_D \exp(-1/[N(0)V]). \qquad (3.14)$$

3.3.4. *Interpretation of Δ_k and E_k defined in Eq. (3.9)*

Upon using Eqs. (3.9) and (3.11), we obtain Eq. (3.7) as

$$\langle\psi_{\text{BCS}}|H_{\text{BCS}}|\psi_{\text{BCS}}\rangle = \sum_k \left(\xi_k - \frac{\xi_k^2}{E_k}\right) - \frac{\Delta^2}{V} \equiv \langle E\rangle_S \qquad (3.15)$$

For the normal state $\Delta = 0$, whence it is found that

$$\langle\psi_N|H_{\text{BCS}}|\psi_N\rangle = \sum_{|k|<k_F} 2\xi_k \equiv \langle E\rangle_N.$$

The energy-difference between the superconducting and the normal states may be written as

$$\langle E_s\rangle - \langle E_N\rangle = \sum_{|k|>k_F} \left(\xi_k - \frac{\xi_k^2}{E_k}\right) + \sum_{|k|<k_F} \left(-\xi_k - \frac{\xi_k^2}{E_k}\right) - \frac{\Delta^2}{V}$$

$$= 2\sum_{|k|>k_F} \left(\xi_k - \frac{\xi_k^2}{E_k}\right) - \frac{\Delta^2}{V}.$$

Replacing the summation in this expression by integration leads to

$$\langle E_s\rangle - \langle E_N\rangle = -\frac{1}{2}\Delta^2 N(0), \qquad (3.16)$$

which is called the condensation energy: It is the amount by which the ground state energy is lowered in the transition from the normal to the superconducting state. It may be used to calculate the critical magnetic field of an SC, which will be further dealt with in Chapter 7.

We now deal with the interpretation of $\Delta_{\mathbf{k}}$ and $E_{\mathbf{k}}$. To this end the expression for $\langle E \rangle_S$ in Eq. (3.15) is written by using Eq. (3.10) as

$$\langle E \rangle_S = -2 \sum_k E_k v_k^4.$$

Consider now the first excited state $\langle E \rangle_{S1}$ which is obtained from this state by dissociation of an electron from the paired state of $(\mathbf{k}, -\mathbf{k})$ to a *definite* free state $-\mathbf{k}'$ for one of the electrons. This implies that $v_{\mathbf{k}'}^2 = 1$. Physical interpretation of $\Delta_{\mathbf{k}}$ and $E_{\mathbf{k}}$ emerges from the following equation:

$$\Delta E = \langle E \rangle_{S1} - \langle E \rangle_S = 2E_{k'} v_{k'}^4 = 2E_{k'} = 2(\xi_{k'}^2 + \Delta^2)^{1/2},$$

where ξ_k is the kinetic energy of an electron scattered out of a CP; it can be arbitrarily small. Hence

$$\Delta E_{\min} = 2\Delta.$$

Therefore 2Δ signifies the minimum energy by which the BCS ground state energy is separated from the energy levels of unpaired electrons. The excitations that have energy $2E_{\mathbf{k}} = 2(\xi_{\mathbf{k}}^2 + \Delta^2)^{1/2}$ are called quasi-particles because the b^+ operators which create them are a strange mixture of single particle operators.

3.3.5. *Temperature-generalization of BCS theory*

The $T = 0$ gap Eq. (3.13a) can now be temperature-generalized because $E_{\mathbf{k}} = (\xi_{\mathbf{k}}^2 + \Delta^2)^{1/2}$ has been identified as the excitation energy a fermion quasi-particle. The probability of its excitation at any temperature T is given by the Fermi function

$$f(E_k) = \frac{1}{\exp(\beta E_k) + 1}, \quad (\beta = 1/k_B T).$$

Since excitation of a state $E_{\mathbf{k}}$ causes the paired state from which it originates to become *non-occupied*, Eq. (3.13a) is temperature-generalized as below

$$1 = [N(0)V] \int_0^{k_B\theta_D} d\xi \frac{1}{(\xi^2 + \Delta^2)^{1/2}} \left[1 - 2f(\beta\sqrt{\xi^2 + \Delta^2}) \right],$$

where the factor of 2 multiplying the Fermi function is included because either of states k or −k may be non-occupied.

Since

$$1 - 2f(x) = 1 - \frac{2}{\exp(x) + 1} = \frac{\exp(x) - 1}{\exp(x) + 1} = \tanh(x/2),$$

the equation for $\Delta(T)$ is obtained as

$$1 = [N(0)V] \int_0^{k_B\theta_D} d\xi \frac{\tanh[(\beta/2)\sqrt{\xi^2 + \Delta^2}]}{(\xi^2 + \Delta^2)^{1/2}}. \tag{3.17}$$

This equation yields two important results.

(i) The equation for T_c, which is defined as the lowest temperature corresponding to which $\Delta = 0$, as:

$$1 = [N(0)V] \int_0^{k_B\theta_D} d\xi \frac{\tanh(\beta_c\xi/2)}{\xi}, \quad (\beta_c = 1/k_BT_c). \tag{3.18}$$

(ii) The equation for $T = 0$ gap Δ_0 as:

$$1 = [N(0)V] \int_0^{k_B\theta_D} d\xi \frac{1}{(\xi^2 + \Delta^2)^{1/2}},$$

or

$$\Delta_0 = \frac{k_B\theta_D}{\sinh(1/[N(0)V])}. \tag{3.19}$$

Therefore, when $[N(0)V] \to 0$, we have

$$\Delta_0 = 2k_B\theta_D \exp(-1/[N(0)V]). \tag{3.20}$$

3.4.   Cooper Problem and BCS Theory Revisited via BSE-based Approach

3.4.1.   *Cooper problem*

This problem is addressed via Eq. (2.14 a):

$$S_{++}(\mathbf{p}) = (-1) \int_L^U d^3\mathbf{q} \frac{\tanh[(\beta/2)\{\Sigma(\mathbf{q}) - E/2\}]}{2\Sigma(\mathbf{q}) - E} I(\mathbf{q} - \mathbf{p})S_{++}(\mathbf{q}),$$

(3.21)

which, via the definition $S_{++}(\mathbf{p})/[2\Sigma(\mathbf{p}) - E] = \varphi(\mathbf{p})$, is written as

$$\varphi(\mathbf{p}) = \frac{-1}{[2\Sigma(\mathbf{p}) - E]} \int_L^U d^3\mathbf{q}\,\tanh[(\beta/2)\{\Sigma(\mathbf{q}) - E/2\}]I(\mathbf{q} - \mathbf{p})\varphi(\mathbf{q}).$$

(3.22)

In this equation $\Sigma(\mathbf{p})$ is the energy of a free electron: $\Sigma(\mathbf{p}) = (\mathbf{p}^2 + m^2)^{1/2}$ and E the total energy of a pair in the centre-of-mass frame. To customize this equation to the situation when the two electrons being considered are at the top of the *non-interacting* Fermi sea at T = 0, we set:

$$\Sigma(\mathbf{p}) = p^2/2m,$$

$$E = 2E_F + W,$$

$$I(\mathbf{q} - \mathbf{p}) = -V/(2\pi)^3 \quad \left(\text{for } E_F \leq \frac{\mathbf{q}^2}{2m},\ \frac{\mathbf{p}^2}{2m} \leq E_F + k_B\theta_D\right)$$

$$= 0, \quad \text{otherwise.}$$

(3.23)

The first of the above equations follows from the fact that the energy of an electron in a free Fermi gas is synonymous with its kinetic energy. Since $\tanh[\ldots] = 1$ at $T = 0$ ($\beta = \infty$), we now obtain Eq. (3.21) as:

$$\varphi(\mathbf{p}) = \frac{V}{(2\pi)^3} \frac{1}{(\mathbf{p}^2/m - 2E_F - W)} \int_{E_F}^{E_F + k_B\theta_D} d^3\mathbf{q}\varphi(\mathbf{q}).$$

Multiplying the above equation with $d^3\mathbf{p}$ and integrating between the limits E_F and $E_F + k_B\theta_D$, we obtain

$$1 = \frac{V}{(2\pi)^3} \int_{E_F}^{E_F + k_B\theta_D} \frac{d^3p}{[\mathbf{p}^2/m - 2E_F - W]}.$$

With $\xi = p^2/2m - E_F$ and $d^3 p = 4\pi p^2 dp = 2\pi(2m)^{3/2}E_F^{1/2}(1+\xi/E_F)^{1/2}$, this is transformed to

$$1 = [N(0)V]\int_0^{k_B\theta_D}\frac{d\xi}{2\xi - W}, \quad \left(N(0) = \frac{(2m)^{3/2}E_F^{1/2}}{4\pi^2}\right),$$

where it assumed that $k_B\theta_D/E_F \ll 1$. Solution of this equation for $[N(0)V] \ll 1$ yields the celebrated Cooper result given earlier in Eq. (3.4):

$$W = -2k_B\theta_D \exp(-1/[N(0)V]).$$

3.4.2. The equation for T_c

This equation is also obtained from Eq. (3.21) by retaining in it the tanh-term which contains β (i.e., $1/k_BT$), employing the definitions for $\Sigma(\mathbf{p})$, E, and $I(\mathbf{q}-\mathbf{p})$ given in Eq. (3.23) *except* that — as was also done in BCS theory — $I(\mathbf{q} - \mathbf{p})$ is now assumed to be non-zero in the range $E_F - k_B\theta_D \leq \mathbf{q}^2/2m$, $\mathbf{p}^2/2m \leq E_F - k_B\theta_D$, rather than in the range $E_F \leq \mathbf{q}^2/2m$, $\mathbf{p}^2/2m \leq E_F + k_B\theta_D$. Therefore with the definition

$$\frac{\tanh[(\beta/2)(\mathbf{p}^2/2m - E_F - W/2)]}{\mathbf{p}^2/2m - E_F - W/2}S_{++}(\mathbf{p}) = \varphi(\mathbf{p}),$$

we have Eq. (3.21) as

$$\varphi(\mathbf{p}) = \frac{V}{(2\pi)^3}\frac{1}{2}\frac{\tanh[(\beta/2)(\mathbf{p}^2/2m - E_F - W/2)]}{\mathbf{p}^2/2m - E_F - W/2}\int_{E_F-k_B\theta_D}^{E_F+k_B\theta_D}d^3\mathbf{q}\varphi(\mathbf{q}).$$

Multiplying the above equation with $d^3\mathbf{p}$ and integrating between the limits $E_F - k_B\theta_D$ and $E_F + k_B\theta_D$, we obtain

$$1 = \frac{V}{(2\pi)^3}\frac{1}{2}\int_{E_F-k_B\theta_D}^{E_F+k_B\theta_D}d^3\mathbf{p}\frac{\tanh[(\beta/2)(\mathbf{p}^2/2m - E_F - W/2)]}{\mathbf{p}^2/2m - E_F - W/2}. \quad (3.24)$$

The equation for T_c follows from this equation by putting W = 0. Since $\tanh(\alpha x)/x$ is even function of x, with $\xi = p^2/2m - E_F$ we now obtain

$$1 = [N(0)V]\int_0^{k_B\theta_D}d\xi\frac{\tanh(\beta_c\xi/2)}{\xi}, \quad (\beta_c = 1/k_BT_c)$$

This equation is identical with Eq. (3.18).

3.4.3. *An equation equivalent to the BCS equation for* $\Delta(T)$

In order to derive an equation for $W(T)$, which will be shown to be equivalent to $\Delta(T)$, we begin once more with Eq. (3.21) and obtain Eq. (3.24). In terms of $\xi = p^2/2m - E_F$, we then have

$$1 = \frac{[N(0)V]}{2}[I_1 + I_2], \tag{3.25}$$

where

$$I_1 = \int_{-k_B\theta_D}^{0} d\xi \frac{\tanh[(\beta/2)(\xi - W/2)]}{\xi - W/2},$$

$$I_2 = \int_{0}^{k_B\theta_D} d\xi \frac{\tanh[(\beta/2)(\xi - W/2)]}{\xi - W/2}.$$

The integral in Eq. (3.25) has been split into two parts because there is a change in the signature of many properties of the system (e.g., velocity of the quasi-particle) upon crossing the Fermi surface (Rickayzen,[1] p. 102). Hence we now assume that

$$W = -|W| \quad for \; \xi > 0, \quad W = |W| \quad \xi < 0. \tag{3.26}$$

The rationale for the first of these equations, which corresponds to the region above the Fermi surface, is provided partly by Cooper problem where too the total energy of the pair was assumed to be $2E_F + W$, and W was found to be negative. Hence we have

$$I_2 = \int_{0}^{k_B\theta_D} d\xi \frac{\tanh[(\beta/2)(\xi + |W|/2)]}{\xi + |W|/2} = \int_{|W|/2}^{k_B\theta_D+|W|/2} dx \frac{\tanh[(\beta/2)x]}{x},$$

and

$$I_1 = \int_{-k_B\theta_D}^{0} d\xi \frac{\tanh[(\beta/2)(\xi - |W|/2)]}{\xi - |W|/2}$$

$$= \int_{|W|/2}^{k_B\theta_D+|W|/2} dx \frac{\tanh[(\beta/2)x]}{x} = I_2,$$

where $\xi - |W|/2 = -x$ has been used for I_1. The equality $I_1 = I_2$ signifies that electron–electron scatterings above the Fermi surface and hole–hole below make equal contributions to the amplitude for pairing.

Substituting the above results into Eq. (3.25), we obtain

$$1 = [N(0)V] \int_{|W|/2}^{k_B\theta_D+|W|/2} dx \frac{\tanh[(\beta/2)x]}{x}. \tag{3.27}$$

In the $T = 0$ limit ($\beta = \infty$), Eq. (3.27) reduces to

$$1 = [N(0)V] \ln \left[1 + \frac{2k_B\theta_D}{|W_0|} \right]. \tag{3.28}$$

Therefore

$$|W_0| = \frac{2k_B\theta_D}{\exp(1/[N(0)V]) - 1},$$

which, in the limit $[N(0)V] \to 0$, reduces to

$$|W_0| = 2k_B\theta_D \exp(-1/[N(0)V]).$$

In the same limit, this expression is identical with the expression for Δ_0 given in Eq. (3.20). In an appendix to this chapter, we will numerically solve Eq. (3.17) for $\Delta(T)$ and Eq. (3.27) for $|W(T)|$ to establish that the latter equation is a viable alternative to the former equation for realistic, i.e., non-vanishing, values of $[N(0)V]$.

3.4.4. *Implications of Eqs. (2.14 b) and (2.14 c) obtained in the previous chapter*

All the results in the previous three sections were obtained via Eq. (2.14 a), which is *one* of the four BSEs that were obtained in the previous chapter. To bring out the implications of Eqs. (2.14 b) and (2.14 c), it suffices to consider the sum of the two tanh-functions that occur in Eq. (2.14 b):

$$\tanh[(\beta/2)\{E/2 - \Sigma(\mathbf{q})\}] + \tanh[(\beta/2)\{E/2 + \Sigma(\mathbf{q})\}]. \tag{3.29}$$

Since

$$\tanh(x) + \tanh(y) = \frac{\sinh(x + y)}{\cosh(x)\cosh(y)},$$

it follows that the Expression (3.29) = 0 when $\beta = \infty$ ($T = 0$), which is also the case for Eq. (2.14 c). Therefore $S_{+-} = S_{-+} = 0$, i.e., the amplitudes for both electron–hole and hole–electron scatterings vanish when $T = 0$.

3.4.5.   *Implications of Eq. (2.14 d) obtained in the previous chapter*

Starting with Eq. (2.14 d), one can obtain an equation of the form of Eq. (3.25) by repeating the steps that led to it starting from Eq. (2.14 a); I_1 and I_2 are now given by

$$I_1 = \int_{-k_B\theta_D}^{0} d\xi \frac{\tanh[(\beta/2)(\xi + W/2)]}{\xi + W/2},$$

$$I_2 = \int_{0}^{k_B\theta_D} d\xi \frac{\tanh[(\beta/2)(\xi + W/2)]}{\xi + W/2}. \tag{3.30}$$

Because these equations have been obtained via the operation of negative-energy projection operators, the signatures of W now must be opposite to those assumed for it in Eq. (3.26) which correspond to the equation obtained via positive-energy operators. Hence

$$W = |W| \quad \text{for } \xi > 0, \quad W = -|W| \quad \xi < 0. \tag{3.31}$$

Employing these signatures for W in Eq. (3.29), we find that $I_1 = I_2$, whence we are led once more to Eq. (3.27). It follows that Eqs. (2.14 a) and (2.14 d) are equivalent equations.

3.5.   Remarks

1. A matter of units and dimensions. $[N(0)V]$ above is dimensionless because the dimensions of $N(0)$ and V in BCS theory are eV^{-1} cm^{-3} and eV cm^3, respectively. In the BSE-based approach, V has been used for $I(\mathbf{q} - \mathbf{p})$, which was introduced to denote the general form of the propagator $D_F(\mathbf{q} - \mathbf{k})$; see Eq. (1.11). Therefore it would seem that we have regarded V as having dimensions of eV^{-2}, as is required of a propagator. This would be incorrect because it overlooks the fact that we are using $\hbar = c = 1$. To elaborate, V in our work continues to have the dimensions of eV cm^3 as in BCS theory, but it comes multiplied with the factor $(\hbar c)^{-3}$. Therefore, with the factors of $\hbar$ and c inserted, we have $[N(0)V]$ as

$$\frac{(2mc^2)^{3/2} E_F^{1/2}}{4\pi^2} \frac{V}{(\hbar c)^3} \Rightarrow \text{dimensionless.}$$

2. Equations (2.14 a) and (2.14 d) were obtained via the operation of positive and negative energy projection operators, respectively. However, given the range of operation of the BCS model interaction in Eq. (3.12), which includes regions both below and above the Fermi surface, we needed to extend the range of applicability of these equations. We have shown that this can be achieved via Eq. (3.26) or (3.30), whence either of Eqs. (2.14 a) and (2.14 d) may be employed to address the superconductivity of elemental SCs. In view of this we regard Eq. (2.14 a) as our basic equation. We now rewrite its equivalent equation, Eq. (3.22), by using the first two equations listed as Eq. (3.23), and defining

$$\tanh[(\beta/2)\{\mathbf{q}^2/2m - E_F - W/2\}] \, \varphi(\mathbf{q}) = \psi(\mathbf{q}).$$

We are thus led to

$$\psi(\mathbf{p}) = (-)\frac{\tanh[(\beta/2)\{\mathbf{p}^2/2m - E_F - W/2\}]}{2(\mathbf{p}^2/2m - E_F - W/2)} \int_L^U d^3\mathbf{q} I(\mathbf{q} - \mathbf{p})\psi(\mathbf{q}).$$

$$(3.32)$$

It will be shown in the next chapter as to how this equation can be adapted to deal with composite SCs.

3. We now draw attention to an equation which follows from Eq. (3.25) and has also been obtained by following a different approach in the books[3] by Schrieffer and by Abrikosov *et al.* Valid at $T = 0$, it is obtained from Eq. (3.25) by naively replacing the tanh-terms in it by (-1) in I_1 and $(+1)$ in I_2. One then obtains

$$\frac{1}{[N(0)V]} = \int_{-k_B\theta_D}^{0} d\xi \frac{(-1)}{2\xi - W} + \int_{0}^{k_B\theta_D} d\xi \frac{(+1)}{2\xi - W}.$$

This is Eq. (7.7) in the book by Schrieffer. Note that solution of this equation *without the first term* yields the result that was obtained in the Cooper problem, which signifies that exclusion of the electrons below the Fermi surface is tantamount to dealing with a 2-electron problem. If the above equation is solved by retaining both the terms, but without changing the signatures of W, i.e., *without imposing* Eq. (3.26), then in the limit of small $[N(0)V]$ one obtains

$$W \simeq \pm i2k_B\theta_D \exp(-1/[N(0)V]).$$

On this basis it has sometimes been said that while Cooper problem is *suggestive* of how the superconducting state may arise, it is "incorrect" because when it takes into account the contribution of holes, then it leads to imaginary or unphysical solutions as given above. This seems to us as an unfair remark because it ignores the difference in the physics of the electron and the hole sectors spelt out in Eq. (3.26); see also (5) below.

4. The role of e–e, e–h, h–e, and h–h scatterings, where e denotes an electron and h a hole, in the original formulation of BCS theory has been a matter of debate in some quarters. While it has implicitly been believed that (a) e–e and h–h scatterings contribute equal to the amplitudes for pairing and (b) e–h and h–e scatterings make no contribution, an explicit demonstration of these features — as provided above — has been lacking in the older literature on superconductivity. However, see also (10) below.

5. We have shown in the previous section that the T-generalized BSE Eq. (2.14 a) leads in the $T = 0$ limit to the expression for the binding energy in the Cooper problem. Further, when $T \neq 0$ and the range of operation of the interaction V is enlarged, the *same* equation has been shown to lead to the BCS expression for T_c and an equation for $|W(T)|$, which is equivalent $\Delta(T)$. Note also that, unlike in the conventional approach, BSE-based approach does not introduce $\Delta(T)$ via an esoteric expression. The latter approach is formulated directly in terms of the more familiar and intuitive concept of the binding energy of the bound state. However, binding energy, i.e., the energy to break up a pair, is now $2|W|$, rather than $|W|$ as is the case when BSE is used for the pairing problem in particle physics. This is so because of the Pauli-blocking of states: for an electron to break off from an initial state $(\mathbf{k}, -\mathbf{k})$ to a final state $\mathbf{k}'$, the latter state has to be available; if not, the pair will not dissociate. This is the reason that energy gap is 2Δ and not Δ.

In so far as BCS theory simply extends the range of application of the model employed by Cooper and incorporates temperature into the theory — which are features, as we have shown, that can easily be accommodated via an extension of the framework of the Cooper problem, one may also view BCS theory simply as an extended form of the theory of Cooper problem. This of course is a statement made with hind sight to emphasize the importance of "language" (BSE in this case) one adopts for the description of any physical phenomenon.

6. Indeed, above simplifications are consequences of the mean-field approximation (MFA) which treats the interaction V as having a constant value because the region of pair-formation, $\pm k_B \theta_D$ around E_F, is very small: $k_B \theta_D / E_F \ll 1$. We should like to note that while attempts have been made to go beyond MFA in *numerous* papers, tangible results to this date have been obtained, more often than not, by *finally* reverting to it. While less ambitious, this seems to us to be a realistic manner to deal with systems that have such huge numbers of constituents. A pragmatic approach then is to fix V with the input of the experimental value of one of the superconducting parameters into one of the equations, and then to use it to predict another superconducting property of the system via another equation; this is precisely how most of the edifice of BCS theory for elemental SCs is built. We deal below with how this program can also be carried out for high-T_c SCs (HTSCs).

7. It is remarkable that Eq. (2.14 a) can also be adapted for HTSCs. This is achieved by replacing the propagator V in it by a superpropagator. The basic idea here is: since an HTSC is a multi-component material, its lattice may be viewed as consisting of more than one sub-lattice, each of them being characterized by its own Debye temperature; pairing could then be caused via simultaneous exchanges of phonons with more than one sub-lattice, in addition to being caused by each sub-lattice alone. One is enabled to implement this idea via the concept of a superpropagator which is a multi-particle propagator — a generalization of the concept of a single-particle propagator. This program will be carried out in the next chapter.

8. In subsequent chapters it will be shown that Eq. (2.14 a) *also* enables one to obtain *dynamics-based* equations for the critical magnetic fields and critical current densities of both elemental and high-T_c SCs.

9. Having brought out the all-important role that Eq. (2.14 a) plays in our work, it seems interesting to point out that an equation derived from it, Eq. (3.13), has been obtained earlier and is well known as the Thouless criterion of superconductivity. However, Thouless's derivation is based on the t-matrix approach, whence it is not immediately evident as to how it can be generalized to deal with multi-component HTSCs — a generalization that is easily carried out in the propagator-based approach of BSE.

10. We return now to the role of e–e and h–h scatterings in BCS theory and note that it has been discussed in a recent book by Annett,[4] who has remarked that BCS state "treats electron and holes in a relatively unsymmetrical manner." He then gives an argument based on a redefinition of the BCS state as a product of two products of wave functions, one pertaining to below the Fermi surface and the other to above it, to conclude that "In fact, electrons and holes contribute more or less equally." The role of e–h and h–e scatterings seems not to have been addressed in this book.

11. It also seems pertinent to draw attention to an *approximate* equation for Δ_0 given by Annett as

$$1 = \lambda \ln \left[\frac{2\hbar\omega_D}{|\Delta_0|} \right]. \quad (\lambda = [N(0)V], \, \hbar\omega_D = k_B\theta_D). \qquad (3.31)$$

We have obtained the *exact* form of this equation, given in Eq. (3.28) [Identify W_0 in this equation with Δ_0 in Eq. (3.31).] Note that both these equations lead to the result given in Eq. (3.20) in the limit of small $[N(0)V]$.

Notes and References

1. L.N. Cooper, *Phys. Rev.* **104**, 1189 (1956).
2. I. Bardeen, L.N. Cooper and J.R. Schrieffer, *Phys. Rev.* **108**, 1175 (1957).
   Authoritative accounts of BCS theory are given in:
   J.M. Blatt, *Theory of Superconductivity* (Academic Press, New York, 1964);
   G. Rickayzen, *Theory of Superconductivity* (Interscience Publishers, New York, 1965);
   M. Tinkham, *Introduction to Superconductivity* (McGraw Hill, New York, 1975);
   D.T. Tilley and J. Tilley, *Superfluidity and Superconductivity* (Overseas Press, New Delhi, 2005)
3. J.R. Schrieffer, *Theory of Superconductivity* (W.A. Benjamin, Reading, Massachusetts, 1964);
   A.A. Abrikosov, L.P. Gorkov and I.E. Dzyaloshinski, *Methods of Quantum Field Theory in Statistical Physics* (Dover, New York, 1963).
4. J.F. Annett, *Superconductivity, Superfluids and Condensates* (Oxford University Press, Oxford, 2008).
5. Differing in matters of detail, the BSE-based approach given above was presented in:
   (a) G.P. Malik and U. Malik, *Physica B* **336**, 349 (2003);
   (b) G.P. Malik, *Physica C* **468**, 949 (2008);
   (c) G.P. Malik, *Int. J. Mod. Phys. B* **24**, 1159 (2010);
   (d) G.P. Malik and M. de Llano, *J. Mod. Phys. 4 A*, **6** (2013).

Appendix A3: A Comparative Study of BCS Equation for $\Delta(T)$ and BSE for (half) the binding energy $W(T)$ of a CP

1. Denoting $[N(0)V]$ by λ, the equations for $|\Delta(T)|$ and $|W(T)|$ obtained in this chapter as Eqs. (3.17) and (3.27), respectively, are:

$$1 = \lambda \int_0^{k_B\theta_D} d\xi \frac{\tanh\left[\left(\frac{1}{2k_BT}\right)\sqrt{\xi^2 + \Delta^2}\right]}{(\xi^2 + \Delta^2)^{1/2}} \qquad (A.3.1)$$

and

$$1 = \lambda \int_{|W|/2}^{k_B\theta_D + |W|/2} dx \frac{\tanh\left[\left(\frac{1}{2k_BT}\right)x\right]}{x}. \qquad (A.3.2)$$

These equations are easily seen to lead to (i) the same equation for T_c which is defined as the lowest temperature at which $\Delta = 0$ or $W = 0$ as:

$$1 = \lambda \int_0^{k_B\theta_D} dx \frac{\tanh\left[\left(\frac{1}{2k_BT}\right)x\right]}{x} \quad (\beta_c = 1/k_BT_c), \qquad (A.3.3)$$

and (ii) the following solutions for Δ_0 and $|W_0|$ when $T = 0$:

$$\Delta_0 = \frac{k_B\theta_D}{\sinh(1/\lambda)} \qquad (A.3.4)$$

$$|W_0| = \frac{2k_B\theta_D}{\exp(1/\lambda) - 1}. \qquad (A.3.5)$$

In the limit of infinitesimal λ, both Eqs. (A.3.4) and (A.3.5) lead to

$$\Delta_0 = |W_0| \simeq 2k_B\theta_D \exp(-1/[N(0)V]).$$

These results already suggest a connection between Δ_0 and $|W_0|$.

2. We now proceed to obtain solutions of Eqs. (A.3.1) and (A.3.2) for some select elemental SCs for finite values of both λ and $T < T_c$ via the procedure given below.

   (i) Determine values of λ of the SCs with the input of their T_c and θ_D into Eq. (A.3.3).

   (ii) Employing the values of λ obtained in (i), calculate the values of Δ_0 via Eq. (A.3.4) and $|W_0|$ via Eq. (A.3.5) and then of $2\Delta_0/k_BT_c$ and $2|W_0|/k_BT_c$. The latter two values in each case have been compared with the corresponding $2\Delta_0/k_BT_c|_{\text{expt}}$ values in Table A.3.1.

Table A.3.1.   Values of $\lambda(T_c)$ obtained by solving Eq. (A.3.3) with the input of θ_D and T_c. Δ_0 and $|W_0|$ are then obtained by solving Eqs. (A.3.4) and (A.3.5), respectively. The experimental values of θ_D, T_c and $\Delta_0|_{expt}$ are taken from C.P. Poole (*Handbook of Superconductivity*, Academic Press, San Diego, 2000).

SC	θ_D (K)	T_c (K)	$\lambda(T_c)$	Δ_0 (meV), $2\Delta_0/k_BT_c$	$\|W_0\|$ (meV), $2\|W_0\|/k_BT_c$	$\Delta_0\|_{expt}$ (meV), $2\Delta_0/k_BT_c\|_{expt}$
Cd	210	0.42	0.1577	0.064, 3.52	0.064, 3.53	0.072, 3.98
Pb	96	7.2	0.3682	1.099, 3.54	1.17, 3.78	1.33, 4.29
Hg	88	4.15	0.3145	0.632, 3.53	0.658, 3.68	0.824, 4.63
Sn	195	3.72	0.2448	0.566, 3.53	0.575, 3.59	0.593, 3.70
In	108	3.41	0.2792	0.519, 3.53	0.533, 3.63	0.541, 3.68
Tl	79	2.38	0.2756	0.362, 3.53	0.372, 3.62	0.395, 3.85
Nb	276	9.25	0.2840	1.407, 3.53	1.449, 3.64	1.55, 3.89

(iii) Write Eqs. (A.3.1) and (A.3.2) in terms of the normalized variables defined as $t = T/T_c$, $\delta = \Delta/\Delta_0$, and $w = |W|/|W_0|$. This means that, by making the substitutions $T = T_c t$, $\Delta = \Delta_0 \delta$ in Eq. (A.3.1), and $T = T_c t$, $|W| = |W_0| w$ in Eq. (A.3.2), we can solve both these equations in the range $0 < t < 1$ to obtain, for each element, values of δ and w in the range $0 < \delta, w < 1$. The plots of these have been given in Fig. A.3.1.

3. A tenet of BCS theory is that it is the same λ that appears in both the equation for T_c and the equation for Δ_0. Therefore, rather than calculating Δ_0 and $|W_0|$ via the value of $\lambda(T_c)$ as was done above, we can also calculate T_cs via the values of $\lambda(\Delta_0)$ and $\lambda(|W_0|)$ obtained from the following equations that follow from Eqs. (A.3.4) and (A.3.5), respectively:

$$\lambda(\Delta_0) = \frac{1}{\operatorname{arcsin} h(k_B\theta_D/\Delta_0)}, \qquad (A.3.6)$$

$$\lambda(|W_0|) = \frac{1}{\ln(1 + 2k_B\theta_D/|W_0|}. \qquad (A.3.7)$$

The results of this exercise are given in Table A.3.2, where the λ-values are obtained by using for both Δ_0 and $|W_0|$ the experimental value of Δ_0.

4. We now draw attention to the fact that the elements we have dealt with here include the so-called "bad actor" SCs, such as Pb and Hg for which the gap-to-T_c ratio departs considerably from the alleged universal value

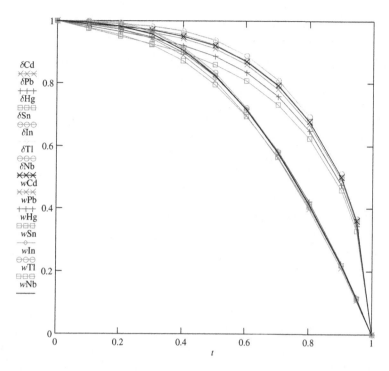

Fig. A.3.1. Plots of $\delta(t)$ and $w(t)$ obtained by solving Eqs. (A.3.1) and (A.3.2) after making the substitutions $T = T_c t$, $\Delta = \Delta_0 \delta$ in Eq. (A.3.1), and $T = T_c t$, $|W| = |W_0|w$ in Eq. (A.3.2). The upper cluster of curves corresponds to $\delta(t)$.

Table A.3.2. Values of T_c calculated via $\lambda(\Delta_0)$ and λ ($|W_0|$) as in Eqs. (A.3.6 and A.3.7).

Element	$T_c|_{\text{expt}}$	$\lambda(\Delta_0)$	T_c via $\lambda(\Delta_0)$	$\lambda(|W_0|)$	T_c via $\lambda(|W_0|)$
Cd	0.42	0.1608	0.47	0.1607	0.47
Pb	7.2	0.3957	8.7	0.3849	8.1
Hg	4.15	0.3430	5.41	0.3372	5.14
Sn	3.72	0.2477	3.9	0.2466	3.83
In	3.41	0.2826	3.56	0.2804	3.46
Tl	2.18	0.2824	2.6	0.2802	2.52
Nb	9.25	0.2920	10.19	0.2894	9.88

of about Eq. 3.53. From the results in the two tables it follows that for *all* the elements dealt with

(i)   The values of $2|W_0|/k_B T_c$ calculated via Eq. (A.3.5) are invariably closer to $2\Delta_0/k_B T_c|_{\text{expt}}$ than the values of $2\Delta_0/k_B T_c$ calculated via Eq. (A.3.4).

(ii) The values of T_c calculated via $\lambda(|W_0|)$ are invariably closer to the experimental values of T_c than the values obtained via $\lambda(\Delta_0)$.

(iii) Invariably, $\lambda(\Delta_0)$, $\lambda(|W_0|) > \lambda(T_c)$. This feature will be further discussed later.

As a consequence of the above, it is seen from Fig. A.3.1 that the cluster of curves corresponding to $w(t)$ is more *compact* than the cluster for $\delta(t)$. Recalling that these clusters correspond not only to the "good" SCs, but also the "bad actor" SCs suggests that compactness of such plots may well be regarded as an alternative criterion of the universality of the theory.

5. We believe that in light of above considerations, Eq. (A.3.2) will be seen to be a more-than-viable alternative to Eq. (A.3.1). More importantly, we note that — as will be shown in the following chapters — the BSE-based approach that leads to Eq. (A.3.2) also enables one to deal with HTSCs via a simple generalization of BCS equations into GBCSEs.

Chapter 4

Generalized BCS Equations for Superconductors Characterized by High-T_cs and Multiple Gaps

---

4.1. Introduction

A liberal[1] estimate for the upper limit of T_c allowed by the weak-coupling BCS theory may be taken as ≈ 23 K, which follows by assuming $\lambda = 0.35$ and $\theta_D = 350$ K. No single-component SC has yet been found with T_c exceeding this limit. In the light of this, a general feature of SCs characterized by high-T_cs and multiple gaps is striking: *all* of them are multi-component materials. This naturally suggests that CPs in these SCs may be bound via simultaneous exchanges of phonons with more than one species of ions — in addition to those that are bound via one-phonon exchanges with each species separately.

It was seen in the previous chapter that BCS theory for elemental SCs can be recast in the framework of BSE where CPs are bound via one-phonon propagators. To address multi-component SCs, we now require a propagator for multiple-phonon exchanges, or a superpropagator.[2] We review in the next section the basic idea of such a field-theoretic construct. In Sec. 4.3, we take up a discussion of another characteristic feature of a composite SC: The anisotropy of its structure which comprises more than one sub-lattice, each of which predominantly contains two species of ions of unequal masses arranged in layers. This feature is addressed via the concept of multiple Debye temperatures (MDTs) — a concept that was first employed by Born and Karmann[3] in the context of a modification of the Debye theory of specific heat of anisotropic solids. The twin concepts of a superpropagator and MDTs are adapted in Sec. 4.4 to obtain generalized BCS equations (GBCSEs). We devote Sec. 4.5 to the Bogoliubov constraint[4] which plays an invaluable role in the application of these equations, and Sec. 4.6 to a general discussion of the procedure to be followed in applying them to any composite SC. Application of the equations to several HTSCs is taken up in the final section.

4.2. Superpropagator

We recall that the Feynman function Δ_F, which describes propagation of the particle associated with a real scalar field $\varphi(x)$ from space-time point x_1 to x_2, is given by the vacuum expectation value of the time-ordered product of $\varphi(x_1)$ and $\varphi(x_2)$:

$$\Delta_F(x_1 - x_2) = \langle 0|T(\varphi(x_1)\varphi(x_2)|0\rangle. \qquad (4.1)$$

If we replace $\varphi(x)$ in this equation by a *superfield* $U\{\varphi(x)\}$, which is a non-polynomial function of $\varphi(x)$, then we obtain the corresponding super-propagator. Some examples of superfields are

$$U\{\varphi(x)\} = \exp\{g\varphi(x)\}, \quad 1/[1 + g\varphi(x)], \quad 1/[1 + g^2\varphi^2(x)], \qquad (4.2)$$

where g is a coupling constant. As an aside we note that non-polynomial theories[2] were originally studied primarily because they were found to have an in-built damping mechanism. It was therefore hoped that they might provide the means of renormalizing, e.g., the theory of weak interactions. They were not pursued any further after the gauge theory of interactions became widely accepted as the correct theory — except that they left the germ of an idea that has been followed up in this monograph. It seems pertinent in this context to note that between the discovery of superconductivity and the emergence of BCS theory, in a period spanning about five decades, many attempts to formulate a theory of the phenomenon proved to be unsuccessful. We draw attention here to a paper by Jörg Schmalian[5] who has made a rather perceptive remark that "(a summary of the unsuccessful attempts to understand superconductivity) illustrates that mistakes are a natural and healthy part of the scientific discourse, and that inapplicable, even incorrect theories can turn out to be interesting and inspiring."

What concerns us here is the final expression for a superpropagator in momentum space and not the rather formidable mathematical exercise which leads to it. For the second superfield in Eq. (4.2), it is given by

$$\Delta_F^{SP}(k) \approx (k^2 g^2)^{-1} \exp(k^2 g^2 / 2) W_{-2, 1/2}(k^2 g^2),$$

where W$_{\lambda, \mu}$ is the Whittaker function. Upon using an appropriate representation of this function, we find that[6]:

$$\Delta_F^{SP}(k) \approx \frac{1}{k^2 g^2} - \int_0^\infty dt \frac{\sigma(t)}{k^2 g^2 + t}, \quad \sigma(t) = e^{-t}(2 - t). \quad (4.3)$$

We now recall from Chapter 1, Eq. (1.8), that the momentum–space transform of Eq. (4.1) is:

$$D(k) = \frac{1}{k^2 - i\varepsilon}. \quad (4.4)$$

A comparison of the last two equations reveals that a superpropagator represents the propagation of a weighted superposition of multiple quanta. It is hence established that a field-theoretic construct exists which may be used in going from the scenario of two particles bound via the exchange of a single particle to the scenario where they are bound via the exchange of multiple particles.

We now make a crucial observation: while we needed CPs to be bound via one-phonon exchange mechanism in recasting BCS theory for elemental SCs in the framework of BSE, we did not use Eq. (4.4) as the kernel of our equation. Rather, appealing to the mean-field approximation — because pairing is assumed to take place in a very narrow region around the Fermi surface — we used in lieu of Eq. (4.4) the *constant* BCS model interaction parameter $-V/(2\pi)^3$ (with appropriate factors of $\hbar$ and c). In the same spirit, in dealing with CPs bound via simultaneous exchanges of phonons with two sub-lattices of the SC, we assume that the kernel of our basic equation obtained in Chapter 2, Eq. (2.14 a), is:

$$I(\mathbf{q} - \mathbf{p}) = -(V_1 + V_2)/(2\pi)^3,$$

$$\left(V_1 = 0, \text{ except for } E_F - k_B\theta_D^{Ac} \leq \frac{\mathbf{q}^2}{2m}, \frac{\mathbf{p}^2}{2m} \leq E_F + k_B\theta_D^{Ac}\right)$$

$$\left(V_2 = 0, \text{ except for } E_F - k_B\theta_D^{Bc} \leq \frac{\mathbf{q}^2}{2m}, \frac{\mathbf{p}^2}{2m} \leq E_F + k_B\theta_D^{Bc}\right), \quad (4.5)$$

where V_1 and V_2 are constants, and θ_D^{Ac} is the Debye temperature of A ions which cause pairing while being *constituents* of a composite — to be distinguished from the Debye temperature θ_D^A in their free state. Similarly, θ_D^{Bc} is different from θ_D^B.

4.3.   Multiple Debye Temperatures

We note that Debye temperature θ_D is just another way to specify Debye frequency ω: $\theta_D = \hbar\omega/k_B$; it is not to be confused with themodynamic temperature. A system at equilibrium is of course characterized by a single value of the thermodynamic temperature. On the other hand, depending upon its constituents, it may be characterized by more than one Debye temperature. If masses of A and B in an anisotropic solid A_xB_{1-x} are different, elastic waves in it travel with different velocities in different directions. This feature causes A and B to have different Debye temperatures, as was noted above. Originally introduced by Born and Karmann, this idea was followed up by Fosterling and others who were thus able to obtain good fits to the experimental results on the molar heats of a number of salts.[3]

We now address the question: Given $\theta_D(x)$ of a binary A_xB_{1-x}, how do we find θ_D^{Ac} and θ_D^{Bc}? We recall[7] that the following equation has often been used to calculate Debye temperature of A_xB_{1-x}, given the Debye temperatures of its constituents:

$$\theta_D(x) = x\theta_D^A + (1 - x)\theta_D^B.$$

Since we are faced with the inverse of this problem, we write this equation as

$$\theta_D(x) = x\theta_D^{Ac} + (1 - x)\theta_D^{Bc}. \quad (4.6)$$

We now require another relation between θ_D^{Ac} and θ_D^{Bc}. In order to meet this requirement, we assume that the modes of vibration of the A and B ions simulate the weakly coupled modes of vibration of a double pendulum. This is suggested by the polariton effect[8] where coupling between the phonon and the photon modes leads to two new modes, different from the original modes of either of them. As shown in Appendix A4, we are thus led to

$$\frac{\theta_D^{Ac}}{\theta_D^{Bc}} = \left[\frac{1 + \sqrt{m_B/(m_A + m_B)}}{1 - \sqrt{m_B/(m_A + m_B)}}\right]^{1/2}, \qquad (4.7)$$

where m_A (m_B) is the atomic mass of an A (B) ion.

4.4. Generalized BCS Equations (GBCSEs) for Composite SCs

Let us first consider the scenario of pairing being caused by two-phonon exchange mechanism (TPEM) in a binary SC A_xB_{1-x}. TPEM implies that there will be CPs bound via simultaneous exchanges of phonons with both the A and the B ions — besides those that are bound via phonon- exchanges with each of them separately. The distinguishing features of the two species of phonons are the values of their net attractive interaction parameters V (as in [N(0)V]) and θ_Ds of A and B.

The equation in the TPEM scenario is obtained from our basic Eq. (2.14 a) which was written as Eq. (3.23) in the previous chapter. After a redefinition of the wave function and using the first two of Eqs. (3.23), it may be rewritten as:

$$\psi(\mathbf{p}) = (-)\frac{\tanh[(\beta/2)\{\mathbf{p}^2/2m - E_F - W/2\}]}{2(\mathbf{p}^2/2m - E_F - W/2)} \int_L^U d^3\mathbf{q} I(\mathbf{q} - \mathbf{p})\psi(\mathbf{q}). \qquad (4.8)$$

Upon substituting for $I(\mathbf{q} - \mathbf{p})$ in this equation the expression given in Eq. (4.5), we obtain

$$\psi(\mathbf{p}) = \frac{\tanh[(\beta/2)\{\mathbf{p}^2/2m - E_F - W/2\}]}{2(\mathbf{p}^2/2m - E_F - W/2)}$$

$$\times \int_{E_F - k_B\theta_D^{Ac}}^{E_F + k_B\theta_D^{Ac}} d^3\mathbf{q}[(V_1 + V_2)/(2\pi)^3]\psi(\mathbf{q}), \qquad (4.9)$$

where (a) we have assumed that $\theta_D^{Ac} > \theta_D^{Bc}$ and (b) the range of integration of the integral multiplying V_2 has been extended from $(E_F - k_B\theta_D^{Bc}, E_F + k_B\theta_D^{Bc})$

to $(E_F - k_B \theta_D^{Ac}, E_F + k_B \theta_D^{Ac})$; this is permissible because, by hypothesis, $V_2 = 0$ in the added range of integration. We now multiply Eq. (4.9) by $\int_{E_F - k_B \theta_D^{Ac}}^{E_F + k_B \theta_D^{Ac}} d^3 \mathbf{p}$ whence

$$1 = \frac{V_1}{(2\pi)^3} \frac{1}{2} \int_{E_F - k_B \theta_D^{Ac}}^{E_F + k_B \theta_D^{Ac}} d^3 \mathbf{p} \frac{\tanh[(\beta/2)\{\mathbf{p}^2/2m - E_F - W/2\}]}{\mathbf{p}^2/2m - E_F - W/2}$$

$$+ \frac{V_2}{(2\pi)^3} \frac{1}{2} \int_{E_F - k_B \theta_D^{Bc}}^{E_F + k_B \theta_D^{Bc}} d^3 \mathbf{p} \frac{\tanh[(\beta/2)\{\mathbf{p}^2/2m - E_F - W/2\}]}{\mathbf{p}^2/2m - E_F - W/2},$$

(4.10)

where we have restored the original limits in the second term. Because W in this equation refers to TPEM, we now denote it by W_2. In terms $\xi = p^2/2m - E_F$, this equation is transformed to

$$1 = [N(0)V_1] \int_{-k_B \theta_D^{Ac}}^{k_B \theta_D^{Ac}} d\xi \frac{\tanh[(\beta/2)\{\xi - W_2/2\}]}{\xi - W_2/2}$$

$$+ [N(0)V_2] \int_{-k_B \theta_D^{Bc}}^{k_B \theta_D^{Bc}} d\xi \frac{\tanh[(\beta/2)\{\xi - W_2/2\}]}{\xi - W_2/2} \left(N(0) = \frac{(2m)^{3/2} E_F^{1/2}}{4\pi^2} \right),$$

(4.11)

where, as usual, $E_F \gg k_B \theta_D^{Ac}$ has been assumed. The equation for T_c in TPEM follows from Eq. (4.11) by putting $W_2 = 0$:

$$1 = \lambda_A^c \int_{-k_B \theta_D^{Ac}}^{k_B \theta_D^{Ac}} d\xi \frac{\tanh(\beta_c \xi/2)}{\xi}$$

$$+ \lambda_B^c \int_{-k_B \theta_D^{Bc}}^{k_B \theta_D^{Bc}} d\xi \frac{\tanh(\beta_c \xi/2)}{\xi}, \quad \lambda_{A,B}^c = [N(0)V_{1,2}]. \quad (4.12)$$

This is manifestly a generalization of the BCS equation for the T_c of an elemental SC. Similarly, following from Eq. (4.11), the equation for W_2 at

$T = 0$, i.e., W_{20}, is

$$1 = \lambda_A^c \ln\left[1 + \frac{2k_B\theta_D^{Ac}}{|W_{20}|}\right] + \lambda_B^c \ln\left[1 + \frac{2k_B\theta_D^{Bc}}{|W_{20}|}\right]. \qquad (4.13)$$

This equation is a generalization of Eq. (3.28) obtained in the one-phonon exchange scenario in the last chapter. Let us denote W_{20} by W_{10} when one of the coupling constants, say λ_B^c, vanishes. We then have (with similar equations when λ_A^c vanishes)

$$1 = \lambda_A^c \ln\left[1 + \frac{2k_B\theta_D^{Ac}}{|W_{10}|}\right],$$

or $\quad |W_{10}^A| = \dfrac{2k_B\theta_D^{Ac}}{\exp(1/\lambda_A^c) - 1} \left(1 = \lambda_A^c \displaystyle\int_{-k_B\theta_D^{Ac}}^{k_B\theta_D^{Ac}} d\xi \dfrac{\tanh(\xi/2k_BT_{c1}^A)}{\xi}\right).$

$$(4.14)$$

The equation within parentheses in Eq. (4.14) is for the T_c at which $|W_{10}^A|$ vanishes. Equations (4.12–4.14) are the GBCSEs in the TPEM scenario that we had set out to obtain. In these equations W_{10} and W_{20} are to be identified with Δ_{10} and Δ_{20}. Note that these three equations contain only two coupling constants. If it is experimentally found that an SC is predominantly characterized by a set of three parameters $\{T_c, \Delta_{20}, \Delta_{10}; \Delta_{20} > \Delta_{10}\}$, then, with the input of, e.g., T_c and Δ_2 into Eqs. (4.12) and (4.13), we can determine these coupling constants. We are thereby enabled to predict the value of Δ_{10} via Eq. (4.14). We follow this approach below because, generally, it is the smaller gap that is known with the least accuracy.

Further generalization of the equations obtained in this section to the situations when CPs are bound via three or more phonon exchange mechanisms is straightforward.

4.5. The Bogoliubov Constraint

Before we proceed to apply GBCSEs to any SC, we need to ask the following question: Is there a principle that guides us as to when the values of λs obtained via Eqs. (4.12) and (4.13) are to be rejected, refined, or accepted? Stated differently, what is the course open to us if one or both of the λs are found to be negative, or greater than unity? The answer to the first question is: Yes and the guiding principle is provided by the Bogoliubov constraint.[4]

It was noted in Chapter 1 that BSE in the ladder approximation is akin to the sum of an infinite geometric series. Since the negative signature of the kernel (V) was explicitly incorporated in our derivation, it is evident that negative values of any of the λs, as well as values greater than unity, must be rejected. The upper limit for λ was investigated by Bogoliubov in a rigorous treatment concerned with renormalization of BCS theory. The value so obtained is[4]:

$$\text{Bogoliubov constraint for BCS theory: } \lambda \leq 0.5. \qquad (4.15)$$

As will be seen below, this constraint plays an invaluable role in the application of GBCSEs to composite SCs.

4.6. Application of GBCSEs to High-T_c SCs: General Considerations

We outline in this section the procedure to be followed in dealing with an SC characterized predominantly by the triplet $\{T_c, \Delta_{20}, \Delta_{10}; \Delta_{20} > \Delta_{10}\}$, which is determined via experiment. The procedure comprises the following steps:

(1) Identify the ion species which may cause pairing in the SC. In the examples below, these are indicated by the entries following the SC:

$$MgB_2: \text{Mg and B,}$$
$$YBa_2Cu_3O_7: \text{Y and Ba,}$$
$$Tl_2Ba_2CaCu_2O_8: \text{Tl, Ba, Ca,}$$

and so on.

(2) Given Debye temperature $\theta_D(x)$ of the composite SC, find via Eqs. (4.6) and (4.7) the Debye temperatures of the ions identified above. In general, this will require taking into account composition of the layer in which the ions identified in (1) are found. In $YBa_2Cu_3O_7$, for example, Ba occurs in the sub-lattice composed of BaO layers. Therefore application of Eqs. (4.6) and (4.7) to this layer will result in two Debye temperatures

for Ba, one when Ba is lower of the two bobs in the double pendulum and the other when it is the upper bob.

(3) For each pair of values of θ_D^{Ac} and θ_D^{Bc} found above, solve Eqs. (4.12) and (4.13) for λ_A^c and λ_B^c with the input of T_c and Δ_{20} of the SC being considered. For the next step below, retain only the pair of $(\lambda_A^c, \lambda_B^c)$ values that is closest to satisfying Eq. (4.15).

(4) If both λ_A^c and λ_B^c selected in the previous step satisfy Eq. (4.15), then use these to calculate two values for the smaller gap via Eq. (4.14). If however one or both of these violate Eq. (4.15), then go back to step (3) and recalculate the λs by a minor fine-tuning of the input values of T_c and Δ_{20}. This is justified because there is a measure of uncertainty in the experimental values of the gaps; also because, unless drop in the property being used to determine T_c is *sharp*, the value of T_c is sometimes quoted as the mean temperature of the range over which the property falls.

(5) After values of λs satisfying Eq. (4.15), or close to satisfying it, have been found go back to the calculation of the smaller gap via Eq. (4.14).

We apply the above procedure to several high-T_c SCs in the next section where the quoted experimental values for various parameters are taken from Poole,[9] except when stated otherwise.

4.7. Application of GBCSEs to Specific High-T$_c$ SCs

4.7.1. MgB_2^{10}

The ion species responsible for TPEM in this SC are Mg and B. We take its $\theta_D(x)$ as 815 K.[9] When Mg is the upper bob in the double pendulum, we have $x = 1/3$. Therefore, with $m(\text{Mg}) = 24.32$ a.m.u. and $m(\text{B}) = 10.81$ a.m.u., Eqs. (4.6) and (4.7) lead to

$$\theta_D^{Mgc} = 1181 \text{ K}, \quad \theta_D^{Bc} = 632 \text{ K}. \tag{4.16}$$

On the hand when Mg is lower of the two bobs, $x = 2/3$, whence we obtain

$$\theta_D^{Mgc} = 322 \text{ K}, \quad \theta_D^{Bc} = 1062 \text{ K}. \tag{4.17}$$

The experimental values of the larger gap and T_c of MgB$_2$ are:

$$\Delta_{20} = 6.2 \text{ meV}, \quad T_c = 39 \text{ K}. \tag{4.18}$$

The inputs of Eqs. (4.16) and (4.18) into Eqs. (4.12) and (4.13) yield

$$\lambda_B^c = 0.1849, \quad \lambda_{Mg}^c = 0.1306. \tag{4.19}$$

Substitution of these together with the corresponding Debye temperatures into Eq. (4.14) leads to the following results: $\Delta_{10}^B = |W_{10}^B| = 0.49$ meV (3.2 K), $\Delta_{10}^{Mg} = |W_{10}^{Mg}| = 0.096$ meV (0.63 K). Since none of these Δ_0-values is close to the experimental value of 2.1 meV, we repeat this exercise with the Debye temperatures given in Eq. (4.17). The results now are:

$$\lambda_B^c = 0.2595, \quad \lambda_{Mg}^c = 0.0492; \quad \Delta_{10}^B = |W_{10}^B| = 3.96 \text{ meV (25.5 K)},$$
$$\Delta_{10}^{Mg} = |W_{10}^{Mg}| \approx 10^{-11} \text{ eV} (\approx 0 \text{ K}).$$

We now carry out a minor fine-tuning of the Δ_{20} value in (4.18) by changing it to 6.28 meV. Equations (4.12) and (4.13) then yield

$$\lambda_B^c = 0.2216, \quad \lambda_{Mg}^c = 0.1073. \tag{4.20}$$

Equation (4.14) now leads to the following results

$$\Delta_{10}^B = |W_{10}^B| = 2.03 \text{ meV (13.1 K)},$$
$$\Delta_{10}^{Mg} = |W_{10}^{Mg}| = 4.975 \times 10^{-6} \text{ eV (0.03 K)}. \tag{4.21}$$

The first of these values is in remarkable agreement with the experimental value of 2.1 meV. In the Remarks section below, we shall discuss the second value.

4.7.2. *YBa$_2$Cu$_3$O$_7$*

The ion species responsible for TPEM in this SC are Y and Ba. Its unit cell comprises layers of

$$\text{BaO/OCu/BaO/CuO}_2\text{/Y/CuO}_2\text{/Y/CuO}_2\text{/BaO/OCu/BaO}.$$

Its $\theta_D(x) = 410$ K, which we assume to be the Debye temperature of each of its sub-lattices. Since the sub-lattice of Y layers contains no other species, $\theta_D^{Yc} = 410$ K. In order to fix θ_D^{Bac}, we first consider a BaO layer with Ba as the *lower* of the two bobs in the double pendulum. Then, with $\theta_D(x) = 410$ K,

$x = 0.5$, $m_2 = m(\text{Ba}) = 137.33$ a.m.u. and $m_1 = m(\text{O}) = 15.999$ a.m.u, Eqs. (4.6) and (4.7) lead to

$$\theta_D^{Bac} = 117 \text{ K}, \quad \theta_D^{Oc} = 703 \text{ K (which is not required).} \quad (4.22)$$

If the positions of the two bobs are reversed, we obtain

$$\theta_D^{Bac} = 478 \text{ K}, \quad \theta_D^{Oc} = 342 \text{ K (which is not required).} \quad (4.23)$$

The experimental values of the larger gap and T_c of $YBa_2Cu_3O_7$ are:

$$\Delta_{20} = |W_{20}| = 20.0 \text{ meV}, \quad T_c = 92 \text{ K.} \quad (4.24)$$

With the input of these and the values of $\theta_D^{Yc} = 410$ K and $\theta_D^{Bac} = 117$ (or 478) K found above, we now solve Eqs. (4.12) and (4.13) to obtain

$$\lambda_Y^c = 0.4147(4.6416), \lambda_{Ba}^c = 0.5354(-3.6839), \quad (4.25)$$

where values in the parentheses correspond to $\theta_D^{Bac} = 478$ K. Since these are in gross disagreement with Bogoliubov constraint Eq. (4.15), we reject them.

For $\theta_D^{Yc} = 410$ K, $\lambda_Y^c = 0.4147$ and $\theta_D^{Bac} = 117$ K, $\lambda_{Ba}^c = 0.5354$, respectively, values of the smaller gap calculated via Eq. (4.14) are

$$\Delta_{10}^Y = |W_{10}^Y| = 7.0 \text{ meV}, \quad \Delta_{10}^{Ba} = |W_{10}^{Ba}| = 3.7 \text{ meV.} \quad (4.26)$$

The experimental value[11] of the smaller gap for this SC is 5.5 meV. It is interesting to note that a minor tuning of the value of $\Delta_{20} = |W_{20}|$ in Eq. (4.24) yields precisely this value. Specifically, the input of $\Delta_{20} = |W_{20}| = 20.4$ meV and $T_c = 92$ K into (4.12) and (4.13) leads to $\lambda_Y^c = 0.3798$ and $\lambda_{Ba}^c = 0.6283$, whence via Eq. (4.14) we obtain

$$\Delta_{10}^Y = |W_{10}^Y| = 5.5 \text{ meV (33.4 K)}, \quad \Delta_{10}^{Ba} = |W_{10}^{Ba}| = 5.2 \text{ meV (27.1 K).} \quad (4.27)$$

By resorting to fine-tuning of the Δ_{20} and/or T_c values of the SC, one could presumably remedy the situation with regard to the value of λ_{Ba}^c, which is now in discord with the Bogoliubov limit. We believe that such an exercise should await more accurate experimental data.

Since we have given a rather detailed account of how GBCSEs are applied to MgB_2 and $YBa_2Cu_3O_7$, the SCs below are dealt with briefly.

4.7.3. $Tl_2Ba_2CaCu_2O_8$

The unit cell of this SC comprises layers of TlO, BaO, CuO_2, and Ca followed by a repetition of these layers. Its $\theta_D(x) = 254\,\text{K}$,[12] which we assume is also the Debye temperature of each of its sub-lattices. Therefore Debye temperatures of the ions of interest, calculated via Eqs. (4.6) and (4.7) with $m(\text{Tl}) = 204.39$ a.m.u. and m(Ba) $= 137.33$ a.m.u., are

$\theta_D^{Cac} = 254$ K (the sub-lattice of Ca layers contains no other species)

$\theta_D^{Tlc} = 289,\,61$ K

> (via TlO layer; second value when Tl is the lower of the two bobs)

$\theta_D^{Bac} = 296,\ 72$ K (via BaO layer;

> second value when Ba is the lower of the two bobs).

Pair-wise, the ion species that may cause TPEM in this SC are Tl and Ba, or Tl and Ca, or Ba and Ca. We quote here the results corresponding to the Ba and Ca ions because this case yields the best results with least "fine-tuning."

The experimental values for this SC are: $T_c = 110$ K[12], $\Delta_{20} = 23.7$ meV.[13] Together with $\theta_D^{Ca} = 254$ K and $\theta_D^{Ba} = 296$ K, if we use $T_c = 110$ K and $\Delta_{20} = 24$ meV as input into Eqs. (4.12) and (4.13), then we obtain $\lambda_{Ba}^c = 0.483$ and $\lambda_{Ca}^c = 0.433$. These values of λ then lead to the following results via Eqs. (4.14):

$$\Delta_{10}^{Ca} = |W_{10}^{Ca}| = 4.8 \text{ meV (33.4 K)}, \quad \Delta_{10}^{Ba} = |W_{10}^{Ba}| = 7.4 \text{ meV (42.3 K)}.$$
$$(4.28)$$

The first of these values is in excellent agreement with the experimental value[13] of 4.7 meV. The second value is a prediction.

4.7.4. $Tl_2Ba_2Ca_2Cu_3O_{10}$

This SC contains the same elements as $Tl_2Ba_2CaCu_2O_8$ from which it differs by having two additional layers, one of Ca and the other of CuO_2. Its $\theta_D(x) = 290$ K,[12] whence

$\theta_D^{Cac} = 290$ K (the sub-lattice of Ca layers contains no other species)

$\theta_D^{Tlc} = 330,\,70$ K

> (via TlO layer; second value when Tl is the lower of the two bobs)

$\theta_D^{Bac} = 338, 83$ K (via BaO layer;

second value when Ba is the lower of the two bobs).

The experimental values for this SC are: $T_c = 115 - 125$ K[13], $\Delta_{20} = 31.0$ meV.[14] In the TPEM scenario via the Ca and Tl ions, the inputs of $\theta_D^{Cac} = 290$ K, $\theta_D^{Tlc} = 330$ K, $T_c = 120$ K, and $\Delta_{20} = 31$ meV into Eqs. (4.12) and (4.13) lead to $\lambda_{Ca}^c = 6.59$ and $\lambda_{Tl}^c = (-)5.11$. The results for the Ca + Ba and Ba + Tl scenarios are similar. If we progressively reduce Δ_{20} in the Ca + Tl scenario, without altering the other parameters, then acceptable values of the two λs appear at $\Delta_{20} = 25.9$ meV:

$$\lambda_{Ca}^c = 0.3738, \quad \lambda_{Tl}^c = 0.5148.$$

These values of λ then lead to the following results via Eq. (4.14):

$$\Delta_{10}^{Ca} = |W_{10}^{Ca}| = 3.7 \text{ meV } (25.8 \text{ K}), \quad \Delta_{10}^{Tl} = |W_{10}^{Tl}| = 9.5 \text{ meV } (53.7 \text{ K}).$$
$$(4.29)$$

The second of these values is close to the experimental value[14] of 10.3 meV.

Our considerations of $Tl_2Ba_2CaCu_2O_8$ suggest that a more detailed investigation of this SC should be taken up when more accurate experimental data are available.

4.7.5. $Bi_2Sr_2CaCu_2O_8$ and $Bi_2Sr_2Ca_2Cu_3O_{10}$

The experimental values for the superconducting parameters of $Bi_2Sr_2CaCu_2O_8$ are given below.

$$\{T_c(K), \Delta_0(\text{meV})\}: \{95, 38\}, \{86, 28\}, \{62, 18\},$$
$$\theta_D: 237 \text{ K}.$$

These values depend on the level of doping of the SC. We note that (a) while for the other SCs dealt with above, the data were available as $\{T_c, \Delta_{20}, \Delta_{10}; \Delta_{20} > \Delta_{10}\}$, they are now available only as $\{T_c, \Delta_0\}$, and (b) for the application of GBCSEs to any SC, values of all the above parameters are required for the same sample. We note that the values of θ_D are seldom available for the samples for which the $\{T_c, \Delta_0\}$ values are quoted.

With $\theta_D = 237$ K, we obtain via Eqs. (4.6) and (4.7) the following values for Debye temperatures of the ions of interest.

$\theta_D^{Cac} = 237$ K, (the sub-lattice of Ca layers contains no other species)

$\theta_D^{Bic} = 269(57)$ K

(via BiO layer; second value when Bi is the upper of the two bobs)

$\theta_D^{Src} = 286(81)$ K (via SrO layer;

second value when Sr is the upper of the two bobs).

These θ_D values enable us to calculate in the TPEM scenario the *maximum* value one might expect for Δ_0 of the SC by solving Eq. (4.13) with the input of (i)$\theta_D^{Src} = 286$ K, $\theta_D^{Bic} = 269$, and (ii) $\lambda_{Bi}^c = \lambda_{Sr}^c = 0.5$. We find that $|W_{20}| = 27.8$ meV. Thus consideration of the set $\{95$ K, **38** meV$\}$ in the TPEM scenario is ruled out. If we now attempt to solve Eqs. (4.12) and (4.13) for the set $\{86$ K, 28 meV$\}$ with $\theta_D^{Cac} = 237$ K and $\theta_D^{Bic} = 269$ K, we find that

(i) $\lambda_{Ca}^c = 25.9$ and $\lambda_{Bi}^c = -22.9$. Since these are in conflict with constraint Eq. (4.15), we need to fine-tune the T_c or the Δ_0 value. Keeping T_c fixed at 86 K and progressively decreasing Δ_0 we find acceptable solutions as $\lambda_{Ca}^c = 0.4555$, $\lambda_{Bi}^c = 0.3614$ at $\Delta_0 = 18.0$ meV, Upon using each of these λ-values along with the corresponding value of θ_D in Eq. (4.14), we obtain

$$\Delta_{10}^{Ca} = |W_{10}^{Ca}| = 5.1 \text{ meV (29.9 K)},$$
$$\Delta_{10}^{Bi} = |W_{10}^{Bi}| = 3.1 \text{ meV (19.2 K)}. \tag{4.30}$$

(ii) Refinement of the unacceptable λ-values in (i) can be also be carried out by keeping Δ_0 fixed at 28 meV and varying T_c. By progressively increasing T_c, we now find nearly acceptable solutions at $T = 121.8$ K as $\lambda_{Ca}^c = 0.5264$, $\lambda_{Bi}^c = 0.5389$. Upon using each of these along with the corresponding value of θ_D in Eq. (4.14), we obtain

$$\Delta_{10}^{Ca} = |W_{10}^{Ca}| = 7.2 \text{ meV (40.2 K)},$$
$$\Delta_{10}^{Bi} = |W_{10}^{Bi}| = 8.6 \text{ meV (47.7 K)}. \tag{4.31}$$

One can only draw a qualitative conclusion from the above calculations because of the assumption made about the value of θ_D (237 K), as also because of the matter of definition of T_c as was discussed earlier. Such a conclusion is: The resistance versus T plot for the SC should not only display a discontinuity at 86 K, but also additional discontinuities.

We turn now to $Bi_2Sr_2Ca_2Cu_3O_{10}$. The experimental values for its superconducting parameters are

$$(T_c(K),\ \Delta_0(meV)\}: \{105, 33\},$$

$$\theta_D: 275\ K^{15}$$

The values of the Debye temperatures of the ions of interest calculated found via Eqs. (4.6) and (4.7) are now found to be: $\theta_D^{Cac} = 275$ K, $\theta_D^{Bic} = 312$ K, and $\theta_D^{Src} = 331$ K. The *maximum* value one might expect for Δ_0 of this SC in the TPEM scenario turns out to be about 30 meV. Hence this SC cannot be dealt with in the TPEM scenario. Additional experimental data, e.g., value of another gap, are required to deal with it in the three-phonon exchange scenario.

4.7.6. *Ba_{0.6}K_{0.4}Fe_2As_2*

The experimental values of the T_c of this iron-pnictide SC available in the literature are: 36.5, 37, and 38 K. The situation with regard to its gaps is extremely fragmentary — both with regard to their number and values. This is so because 2-4 gaps have been reported for it with values in meV as: 6, 12; 2.5, 9.0; 3.3, 7.6; 3.6, 8.5, and 9.2; 4, 7, and 12 with an additional gap at 9.5 (original references to these values can be found in the last paper cited under Ref. 16 in Notes and References below).

For the sake of concreteness, we take up here the task of addressing via GBCSEs in the TPEM scenario the following data reported by Shan *et al.*[16]

$$\{T_c, \Delta_{20}, \Delta_{10}\} = \{38\ K, 8.3\ meV, 3.05\ meV\}. \tag{4.32}$$

Pairing in this SC via TPEM can be caused by the Ba and the Fe ions, as well as by the Ba and the As ions. We confine ourselves here to the Ba + Fe scenario. Given Debye temperature of the SC as 274 K, the relevant Debye temperatures of these ions calculated via Eqs. (4.6) and (4.7) are:

$$\theta_D^{Bac} = 125\ K, \quad \theta_D^{Fec} = 399\ K. \tag{4.33}$$

The solution of Eqs. (4.12) and (4.13) with the input of these θ_Ds, together with the values of T_c and Δ_{20} from Eq. (4.32), yields $\lambda_{Ba}^c = 1.263$ and $\lambda_{Fe}^c = (-)0.276$. Since these values are in disagreement with Eq. (4.15), we resort to fine-tuning of Δ_{20}. Upon progressively decreasing it, we find the following solutions at $\Delta_{20} = 7.3$ meV

$$\lambda_{Ba}^c = 0.4818, \quad \lambda_{Fe}^c = 0.1442. \tag{4.34}$$

Upon now solving Eq. (4.15), we obtain

$$\Delta_{10}^{Ba} = |W_{10}^{Ba}| = 3.1 \text{ meV (17.8 K)},$$
$$\Delta_{10}^{Fe} = |W_{10}^{Fe}| = 6.7 \times 10^{-5} \text{ eV (0.4 K)}. \tag{4.35}$$

Hence it is seen that θ_D-values in Eq. (4.33) and λ-values in Eq. (4.34) yield for the set $\{T_c, \Delta_{20}, \Delta_{10}\}$ the values $\{38 \text{ K}, 7.3 \text{ meV}, 3.1 \text{ meV}\}$, which are in good agreement with the experimental values if the uncertainties in their determination are taken into account.

4.8. Remarks

1. It was shown in the previous chapter that BCS equations for an elemental SC obtained via a variational approach may also be obtained via an alternative approach based on a T-generalized BSE, employing as its kernel the one-phonon propagator. One may view CPs in the latter approach as being bound together via the "glue" provided by a single spring. It then follows that when a superpropagator is employed as the kernel of the BSE, CPs will be bound together by the stronger glue provided by a composite spring. It is this feature that causes an enhancement of T_c of a non-elemental SC.

2. Re: Isotope effect in HTSCs. BCS theory gives the value of α as 0.5 in the relation

$$T_c \propto M^{-\alpha}, \tag{4.36}$$

where M is the average mass of the ions in the lattice of a single element. Relation Eq. (4.36) is a statement of the Isotope effect. This effect helped in formulation of the theory because it sheds light on the role of the ion lattice for the phonon-mediated mechanism for pairing. We note however that values of α significantly different from 0.5 have also been found for some elements, such as Mo, Os, Ru, and Zr for which it has the values 0.33, 0.2, 0, and 0, respectively. Hence (4.36) does not have the status of a *law*.

The applicability of a relation similar to Eq. (4.36) has also been investigated for HTSCs such as YBa$_2$Cu$_3$O$_7$ (YBCO) and MgB$_2$. For the former it has been reported that $\alpha = 0$, which follows from the experimental finding that replacement of up to 75% of O-16 by O-18 did not lead to any change in the T_c of the SC. Note that Eq. (4.36) is a relation based on phonon exchanges with a single lattice, whereas our treatment of the T_c and the gaps of YBCO was based on phonon exchanges with two distinct sub-lattices — one comprising layers of the Y ions alone, and the other comprising layers of BaO. In the paragraph following Eq. (4.26) it was pointed out that the input of $\Delta_{20} = 20.4$ meV and $T_c = 92$ K in our approach yields

$$\lambda_Y^c = 0.3798(\theta_D^{Yc} = 410 \text{ K}), \quad \lambda_{Ba}^c = 0.6238(\theta_D^{Bac} = 117 \text{ K}),$$

which, in the one-phonon exchange scenario, leads to

$$\Delta_{10}^Y = 5.5 \text{ meV} (33.4 \text{ K}), \quad \Delta_{10}^{Ba} = 5.2 \text{ meV} (27.1 \text{ K}).$$

In the parentheses in the last two equations are given the temperatures at which these gaps vanish (the experimental value of the smaller gap is 5.5 meV). If we assume that that Debye temperature of YBCO remains unchanged when O-16 is replaced by O-18, then Debye temperature of the sub-lattice containing Y ions continues to be 410 K, whereas Debye temperature of the Ba ions in the sub-lattice containing BaO changes from 117 to 122 K — vide Eqs. (4.6) and (4.7). Equations (4.12) and (4.13) then yield (input $\Delta_{20} = 20.4$ meV and $T_c = 92$ K):

$$\lambda_Y^c = 0.3686(\theta_D^{Yc} = 410 \text{ K}), \quad \lambda_{Ba}^c = 0.6333(\theta_D^{Bac} = 122 \text{ K}).$$

In the OPEM scenario, these (λ, θ) values lead to

$$\Delta_{10}^Y = 5.0 \text{ meV} (30.8 \text{ K}), \quad \Delta_{10}^{Ba} = 5.5 \text{ meV} (28.7 \text{ K}).$$

Therefore in looking for the isotope effect — even when it is found that the T_c at which Δ_{20} vanishes does not change — we need to monitor the temperature at which the smaller gap vanishes.

Above considerations bring out that for a realistic study of the isotope effect in MgB$_2$, or any other HTSC, we require not only the experimental values of T_c, Δ_{10}, $\Delta_{20} > \Delta_{10}$, and θ_D — both before and after an isotope is replaced by another, but also the temperature at which Δ_{10} vanishes.

3. Given the values of θ_D and T_c (or Δ_0) of an elemental SC, BCS equations enable one to predict the value of its Δ_0 (or T_c). In principle, GBCSEs play a similar role: given values of any two parameters of the set $\{T_c, \Delta_{20}, \Delta_{10}\}$ of a non-elemental SC, they enable one to predict the third. In practice, however, one has to resort to fine-tuning of one or the other parameter because of the uncertainty in its experimental value.

4. For every SC dealt with above, we had to resort to fine-tuning of one or the other parameter because of the uncertainty in its experimental value. It therefore seems imperative that we discuss in some detail as to how fine-tuning is done and what exactly it accomplishes. We now do so by taking the example of MgB_2. The experimental values of its superconducting parameters were taken as: $T_c = 39$ K, $\Delta_{20} = 6.2$ meV, $\Delta_{10} = 2.1$ meV. Assuming that the value of its T_c is accurate, but not that of Δ_{20}, we now solve Eqs. (4.12) and (4.13) with the input of Debye temperatures from Eq. (4.17), $T_c = 39$ K, and values of Δ_{20} in the range: 6.0 meV $\leq \Delta_{20} \leq$ 6.7 meV. The values of λ_B^c and λ_{Mg}^c so obtained, together with the corresponding values of $|W_{10}| = \Delta_{10}$ (where applicable) obtained by solving Eq. (4.14) are tabulated below.

$|W_{20}| = \Delta_{20}$ (meV)	λ_B^c	λ_{Mg}^c	$|W_{10}^B| = \Delta_{10}^B$ (eV)	$|W_{10}^{Mg}| = \Delta_{10}^{Mg}$ (eV)
6.0	0.3411	−0.076	—	—
6.1	0.3024	−0.0167	—	—
6.2	0.2595	0.0492	3.96×10^{-3}	8.3×10^{-11}
6.3	0.2116	0.1227	1.64×10^{-3}	1.6×10^{-5}
6.4	0.1577	0.2052	3.23×10^{-4}	4.3×10^{-4}
6.5	0.0969	0.2985	6.0×10^{-6}	2.0×10^{-3}
6.6	0.0277	0.4046	3.84×10^{-17}	5.1×10^{-3}
6.7	−0.0518	0.5265	—	—

This exercise brings out that if we start with a particular value of Δ_{20} (with T_c fixed) and are led to values of λs in conflict with constraint Eq. (4.15), then "refinement" of Δ_{20} cannot be carried out indefinitely. Rather, it leads to a window of Δ_{20} values which leads to values of the two λs in agreement with Eq. (4.15). In the Table above, this window is provided by: 6.2 meV $\leq \Delta_{20} \leq$ 6.6 meV. Note that the value of

$\Delta_{20} = 6.5$ meV yields a value for the smaller gap as 2.0 meV, which is also close to the experimental value of 2.1 meV. At this stage one must make a judicious choice: $\Delta_{20} = 6.28$ meV (as was adopted by us in the text above) or 6.5 meV? This is an issue that can be settled by taking into account the error bars on the experimental values of Δ_{20}.

The window that we found for MgB$_2$ is a general feature: Application of Eqs. (4.12) and (4.13) leads to such a window for each of the SCs dealt with above.

5. Since GBCSEs embody the twin concepts of the existence of CPs in a composite SC with different binding energies and MDTs, we consider it pertinent to draw attention below to a few papers where the need for such concepts has been articulated, or evidence for them found via experiment.

(i) Dealing with MgB$_2$ in the strong-coupling Eliashberg formalism, Choi *et al.*[17] *assumed* that it is characterized by two gaps. They could then account for its observed superconducting features by *postulating* the existence of two separate populations of electrons that form pairs with different binding energies on different parts of the Fermi surface.

(ii) In a paper by Liu *et al.*[18] the need for two-phonon scatterings is reflected in the title itself: "Beyond Eliashberg Superconductivity in MgB$_2$: Anharmonicity, Two-phonon Scattering and Multiple Gaps." In this paper the coupling constant is decomposed into contributions from four different bands crossing the Fermi surface in order to analyze the effects of interband anisotropy on the T_c and gap-structure of the SC. They were thus led to $\lambda_{\text{effective}} = 1.01$ and concluded that it is "arguably consistent with the measured T_c of nearly 40 K."

(iii) In dealing with the thermal conductivity of MgB$_2$, Sologubenko *et al.*[19] followed an approach that was summarized by them as: "Thus, we consider two subsystems of quasi-particles with gaps Δ_1 and Δ_2, different parameters E_1 and E_2 of phonon-electron scattering, and separate contributions to (thermal conductivities) κ_{c1} and κ_{c2} to the heat transport." In this manner — in the early days of MgB$_2$ — they were able to give good estimates of the gaps at $T = 0$.

(iv) The title of a paper by Tacon *et al.*[20] based on electronic Raman scattering experiments, "Two energy scales and two distinct quasiparticle dynamics in the superconducting state of underdoped cuprates," *per se* is supportive of the twin concepts under consideration.

(v) Finally, we draw attention here to a paper by Kwei and Lawson[21] where, on the basis of neutron powder diffraction experiments, different Debye temperatures for the constituents of La_2CuO_4 have been reported.

Even from the brief account given above, which is based on a rather random selection of papers, it will be seen that GBCSEs provide a concrete and compact form to concepts that have been articulated from time to time by various authors in the context of high-T_c SCs.

6. T_cs of some of the HTSCs dealt with in this chapter satisfy the following inequalities:

$$T_c(MgB_2) < T_c(YBCO) < T_c(\text{Bi-2212}) < T_c(\text{Bi-2223})$$
$$T_c(\text{Bi-2212}) < T_c(\text{Tl-2212})$$
$$T_c(\text{Bi-2223}) < T_c(\text{Tl-2223}).$$

It is therefore seen that *richer* the structure of the SC, greater is its T_c. *Richness* is meant here to signify (a) the *diversity* of the constituents of the SC — in the sense that that MgB_2 has two kinds of constituents (alkaline Mg and metalloid B), whereas YBCO has three kinds (transition metals Y and Cu, alkaline Ba, and a non-metal O), and (b) the *number of sub-lattices* which could cause pairing of conduction electrons. Following from the above inequalities is also a relevant question: why do replacements of Bi by Tl and Sr by Ba cause enhancement in the T_c of the SC? In order to deal with this question, we assume that pairing in Bi-2212 in the TPEM scenario takes place via the Bi and Sr ions (rather than the Bi and Ca ions considered in Sec. 4.7.5), and via the Tl and Ba ions in Tl-2212 (as an alternative to the Ba and Ca ions considered in Sec. 4.7.3).

With the θ_D and λ values as given in the Table below, Eqs. (4.12) and (4.13) lead to $T_c = 70$ K, $|W_{20}| = 13.5$ meV for Bi-2212, and to $T_c = 110$ K, $|W_{20}| = 23.6$ meV for Tl-2212.

Element X	At. No.	At. mass (amu)	Valence	θ_D^{Xc} in Bi-2212/Tl-2212	λ_X^c in Bi-2212/Tl-2212	Electron affinity (EA) kJ/mol	Electron-egativity (EN) (Pauling scale)
Bi	83	208.98	5	269	0.1613	91.2	2.02
Tl	81	204.39	3	289	0.3566	19.2	1.62
Sr	38	87.62	2	286	0.4943	5.03	0.95
Ba	56	137.33	2	296	0.5172	13.95	0.89

It is seen from the properties of the elements given in the above Table that (a) the θ_D values for Tl-2212 are only marginally greater than those for Bi-2212 and (b) the *sum* of the λ values for the former (Tl + Ba) is 0.8738 and 0.6556 for the latter (Bi + Sr). Hence the question: to what factors do we attribute the increase in the values of λ which cause T_c to increase? Because this question involves a discussion of the Femi surface of the SC and the constituents of λ, we shall deal with it after the next remark below. In so far as the valence-, EA-, and EN- values given in the Table are concerned, they may well be irrelevant for the question being addressed because, among other reasons, they are not the *in-situ* values. Nonetheless we note that in going from Bi-2212 to Tl-2212, there is a decrease in the overall values of both EA and EN — from 96.23 to 33.15 kJ/mol for the former and from 2.97 to 2.51 for the latter.

7. It might seem that our approach to deal with a composite SC on the basis of MDTs and multiple exchanges of phonons between electrons is rather naive because it does not take into account the highly complex structure of its Fermi surface, which comprises many sheets. It is pertinent in this connection to recall that Fermi surfaces of most of the elements too have rather complex structures[22] — *none* of them has a perfectly spherical Fermi surface assumed in the BCS theory. And yet MFA works for them.

We now draw attention to the fact that the *mean* value of E_F when Eqs. (4.14) is used for pairing in the OPEM scenario via the A-ions (or the B-ions) is in general different from the one when Eq. (4.12) or (4.13) is employed in the TPEM scenario for pairing via *both* the A- and the B-ions. This follows from the SC having multiple gaps, which are supposed to follow undulations of the Fermi surface. Thus complexity of structure of a composite SC in our approach is being addressed implicitly by assuming its Fermi surface to be *locally spherical* with different mean values of E_F as their radii. This is not immediately evident because none of these E_Fs is required for the calculation of either the T_c or any of the Δs of the SC — just as it is not required when elemental SCs are dealt with via the usual BCS equations. This is so because of the assumption that even the smallest value of E_F in any of these cases is much greater than the corresponding Debye cut-off energy. We note in passing that locally spherical Fermi surfaces are reminiscent of the *locally inertial co-ordinate frames* in the general theory of relativity; this will be further discussed in Chapter 11.

8. To return to the question concerned with the increase in the T_c of Bi-2212 raised in Remark 6, we first recall the definition of λ given in Eq. (4.11):

$$\lambda \equiv [N(0)V] \sim m^{3/2} E_F^{1/2} V.$$

In these relations, m is the effective mass of an electron and V the net attractive interaction between a pair of electrons due both to attraction because of the ions and the Coulomb repulsion. We next note that all the HTSCs in this chapter were dealt with without invoking the values of any of the constituents of λ; this was so because of the assumptions that (a) as already noted, mean values of E_F are much greater than the Debye cut-off energies and (b) $\lambda(T = 0) = \lambda(T = T_c)$.

We now observe that assumption (a) may well not be valid for Bi-2212. If so, we need equations for the T_c and the Δs of the SC that explicitly incorporate E_F or, strictly speaking, the chemical potential μ. In other words, the increase in the T_c of Bi-2212 when Bi is replaced by Tl and Sr by Ba may well be due to changes in the μ values — and the concomitant changes in other variables — that these substitutions cause. Since μ is related to the density of the charge carriers in the SC, this is an inference that be checked by experiment via, e.g., the Hall effect.

9. As is well known, there are "good" and "bad actor" elemental SCs. To the former category belong *most* of the elemental SCs for which

λ $(T = 0)$ calculated via the BCS equation for Δ_0 nearly equals λ $(T = T_c)$ calculated via the equation for T_c. This is the reason for the relation $2\Delta_0/k_B T_c \simeq 3.53$ to be called a universal relation. For the latter category of SCs, such as Pb, Hg, and Nb, this relation is not satisfied. This observation leads us to suggest that HTSCs may well be similarly characterized. The rationale behind this suggestion is the fact that while we could, with minor fine tuning, account for the observed values of the T_c and the Δs of, e.g., MgB_2 and YBCO, we could not do so for Bi-2212. Hence the statement that different values of the gap-to-T_c ratio reported in the literature for some of the HTSCs comprise knowledge that is "fragmentary" needs a review. To elaborate, it seems to us that all the (T_c, Δ) values noted in Sec. 4.7.5 for Bi-2212 may well be accounted for by equations incorporating μ. Derivation of such equations and their applications will be taken up in Chapters 9–11.

10. We turn now to a discussion of the infinitesimal values of the gaps that we found, e.g., for MgB_2 in Eq. (4.21). In our earlier papers, we had remarked that such gaps are too small to be observed. It turns out however that precisely such gaps have recently been observed[23–25] in the iron-pnictide SCs, where they appear as nodes or line nodes. They are currently evincing a lot of interest.

11. A general feature that emerges when GBCSEs are applied to any composite SC is: on dealing with the set $\{T_c, \Delta_{20}, \Delta_{10}\}$ of any SC via TPEM, one is led to several different values for Δ_{10}, each of which vanishes at a different value of T_c. Consider, for example, $Bi_2Sr_2CaCu_2O_8$. TPEM in this SC can be caused by Bi + Sr, or Bi + Ca, or Sr + Ca, ions. In each of these cases, one is led to two values for Δ_{10}. This implies that the resistance versus temperature plot for this SC should display many discontinuities. Such discontinuities have actually been observed and are displayed in Ref. 15 in a figure which is reproduced (with permission) in Ref. 26(e). While these discontinuities are generally attributed to the presence of more than one phase in the SC, they may well alternatively or additionally be due to breaking up of CPs bound via one-phonon exchange mechanism, before the T_c is reached at which pairs bound via TPEM break up.

12. We conclude this chapter by noting that BCS theory relates the relevant parameters of an elemental SC after it has reached the superconducting

state; it is not a detailed theory of how such a state is reached. GBCSEs play a similar role for non-elemental SCs. In other words, they cannot predict as to how the properties of, say, $Bi_2Sr_2CaCu_2O_{8+x}$, will change with x; given the values of its superconducting parameters for any value of x, they merely provide correlations among them.

Notes and References

1. This is a liberal estimate because, among elements, Nb has the highest $T_c = 9.25$ K. This value follows from $\theta_D = 276$ K and $\lambda = 0.284$.
2. A. Salam (Ed.), *Nonpolynomial Lagrangians, Renormalization and Gravity*, Gordon and Breach, NY, 1971.
3. For a review of the body of work on the molar specific heats based on multiple Debye temperatures, see: F. Seitz, *The Modern Theory of Solids* (McGraw Hill, NY, 1940).
4. For a discussion see: J.M. Blatt, *Theory of Superconductivity* (Academic Press, NY, 1964).
5. Jörg Schmalian, *Failed theories of Superconductivity*, available on the internet.
6. S.N. Biswas, G.P. Malik and E.C.G. Sudarshan, *Phys. Rev. D* **7**, 2884 (1973).
7. R. D. Parks (Ed.), *Superconductivity*, Vols. 1 and 2 (Marcel and Dekker, NY, 1969).
8. M.A. Omar, *Elementary Solid State Physics: Principles and Applications* (Addison-Wesley, Reading Massachusetts, 1975), p. 128.
9. C.P. Poole, Jr., *Handbook of Superconductivity* (Academic Press, San Diego, 2000).
10. Data for this SC are taken from: C. Buzea and T. Yamashita, *Superconductors, Science and Technology* **14**, R115 (2001); the value of 815 K for its Debye temperature is the mean of 750 and 880 K given in this Review.
11. N. Klein *et al.*, *Phys. Rev. Lett.* **71**, 3355 (1993); M.A. Bari *et al.*, *Physica C* **256**, 227 (1996).
12. A. Junod *et al.*, *Physica C* **159**, 215 (1989).
13. J. Gopalakrishnan, in: A.M. Hermann and J.V. Yakhmi (Eds.) *Thallium-Based High-Temperature Superconductors* (Dekker, NY, 1994), p. 13.
14. K.F. Renk, in Ref. 11, p. 507.
15. T. Triscone and A. Junod, in: H. Maeda and K. Togano (Eds.), *Bismuth-Based High Temperature Superconductors* (Academic Press, San Diego, 2000).
16. L. Shan *et al.*, *Nature Phys.* **7**, 325 (2011).
17. H.J. Choi *et al.*, *Nature* **418**, 758 (2002).
18. A.Y. Liu *et al.*, *Phys. Rev. Lett.* **87**, 087005 (2001).
19. A.V. Sologubenko *et al.*, *Phys. Rev. B* **66**, 0145504 (2002).
20. M. Le Tacon *et al.*, *Nature Phys.* **2**, 537 (2006).
21. G.W. Kwei and A.C. Lawson, *Physica C* **175**, 135 (1991).
22. A.P. Cracknell and K.C. Kong, *The Fermi Surface*, Clarendon Press, Oxford, 1973.
23. D. Lee, *Nature Phys.* **8**, 364 (2012).
24. Y. Zhang *et al.*, *Nature Physics* **8**, 371 (2012).
25. M.P. Allan *et al.*, *Science* **336**, 563 (2012).

26. The matter covered in this chapter is based on the papers noted below, where the specific topic each of them dealt with is also indicated. The account of the T_Cs and gaps of SCs given above was rather brief. In the papers below these SCs have been dealt with in far greater detail.

(a) G.P. Malik and U. Malik, *Physica B* **336**, 349 (2003): showed that a superpropagator can cause enhancement in the value of T_C;

(b) G.P. Malik and U. Malik, *Physica B* **348**, 341 (2004): introduced MDTs in the context of superconductivity;

(c) G.P. Malik, *Int. J. Mod. Phys. B* **24**, 1159 (2010): Was concerned with establishing equality between Δ and $|W|$, and obtaining GBCSEs;

(d) G.P. Malik, *Int. J. Mod. Phys. B* **24**, 3701 (2010): dealt with MgB_2 and $YBa_2Cu_3O_7$;

(e) G.P. Malik and U. Malik, *J. Supercond. Nov. Magn.* **24**, 255 (2011): dealt with the Tl- and Bi-based SCs;

(f) G.P. Malik, M. de Llano, and I. Chávez, *J. Mod. Phys.* **4**, 474 (2013): dealt with $Ba_{0.6}K_{0.4}Fe_2As_2$.

We should like to add that the papers above dealing with MgB_2, and the Tl- and Bi-based SCs, respectively, do away with the arbitrariness with which they were earlier dealt with in G.P. Malik, *Physica C*, **433**, 70 (2005) and G.P. Malik, in: B.P. Martins (Ed.), *New Research on Superconductivity* (Nova, NY, 2007), p. 3. These papers were written prior to the equality between Δ with $|W|$ (half the binding energy of a CP) was established.

Appendix A4: Small Oscillations of a Co-Planar Double Pendulum

Thermal oscillations of the two sub-lattices that constitute the lattice of a binary were assumed in the text to simulate the small oscillations of a co-planar double pendulum of equal lengths. We give below derivation of formula given in Eq. (4.7) which is based on Landau and Lifshitz, *Mechanics* (Addison-Wesley, Reading, MA, 1975), pp. 11 and 70.

In the frame shown in Fig. A.7.1, Lagrangian L for the double pendulum is

$$L = \frac{1}{2}l^2(m_1 + m_2)\dot{\varphi}_1^2 + \frac{1}{2}m_2 l^2 \dot{\varphi}_2^2 + m_2 l^2 \cos(\varphi_1 - \varphi_2)\dot{\varphi}_1 \dot{\varphi}_2$$
$$+ (m_1 + m_2)gl\cos\varphi_1 + m_2 gl \cos\varphi_2.$$

For small oscillations ($\varphi_1, \varphi_2 << 1$) this reduces to

$$L_{red} = \frac{1}{2}l^2(m_1 + m_2)\dot{\varphi}_1^2 + \frac{1}{2}m_2 l^2 \dot{\varphi}_2^2 + m_2 l^2 \dot{\varphi}_1 \dot{\varphi}_2$$
$$- \frac{1}{2}(m_1 + m_2)gl\varphi_1^2 - \frac{1}{2}m_2 gl\varphi_2^2.$$

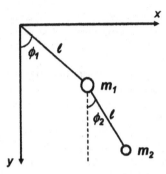

Fig. A.7.1.   A double pendulum.

Therefore the equations of motion are:

$$(m_1 + m_2)l\ddot{\varphi}_1 + m_2 l\ddot{\varphi}_2 + (m_1 + m_2)g\varphi_1 = 0,$$
$$l\ddot{\varphi}_1 + l\ddot{\varphi}_2 + g\varphi_2 = 0.$$

Letting

$$\varphi_1 = A\cos(\omega t + \delta), \qquad \varphi_2 = B\cos(\omega t + \delta),$$

where A, B, ω, and δ are unknown constants, we obtain

$$A(m_1 + m_2)(l\omega^2 - g) + Bm_2 l\omega^2 = 0, \quad Al\omega^2 + B(l\omega^2 - g) = 0.$$

The roots of the characteristic determinant are therefore given by

$$\omega_{1,2}^2 = \frac{g(m_1 + m_2)[1 \pm \sqrt{1 - m_1/(m_1 + m_2)}]}{m_1 l},$$

whence

$$\frac{\omega_1}{\omega_2} = \frac{\theta_1^c}{\theta_2^c} = \left[\frac{1 + \sqrt{1 - m_2/(m_1 + m_2)}}{1 - \sqrt{1 - m_2/(m_1 + m_2)}}\right]^{1/2}.$$

which is relation given in Eq. (4.7) in the text. Note that the above ratio depends upon the relative positions of two bobs of the double pendulum.

Chapter 5

Multi-Gap Superconductivity: Generalized BCS Equations (GBCSEs) as an Alternative to the Approach Due to Suhl, Matthias, and Walker (SMW)

5.1. Introduction

It was shown in the last chapter that GBCSEs enable one to address the T_cs and multiple gaps of HTSCs in the manner elemental SCs are dealt with in the original theory. Because multiple gaps of an SC have routinely been addressed via the widely-known SMW[1] approach, it seems imperative that we juxtapose this approach with the approach provided by GBCSEs. To this end, we first review the SMW approach and draw attention to an important feature of it that seems to have escaped serious attention. This is followed by a study of the T_c and gaps of MgB_2 via both the approaches.

5.2. A Review of the SMW Approach for a Two-gap Superconductor

(i) This approach was originally given in the context of transition elements, a peculiar feature of which is the filling up of $3d$ orbital *after* the $4s$. This implies that valence electrons in these are divided between two bands. Since the d-band is poorly occupied, and since the effective mass of d-electrons is large, it is expected to make negligible contribution to the conductivity of the material. However, the availability of many vacant levels in the d-band also implies that it has a large density of states. Hence the s-electrons can be scattered not only within their own band, but also to the d-band. This is known to have a considerable effect on the conductivity of transition elements in the normal state. The motivation for SMW approach was provided by the question: do $s-d$ transitions have a similar effect in the superconducting state?

(ii) Above considerations led SMW to consider a Hamiltonian comprising three pieces which correspond to electrons in the (a) s-band with N_s as the density of states, (b) d-band with N_d as the density of states, and (c) the overlap between the s- and d-bands. Recalling the Hamiltonian corresponding to the case of a single band, Eq. (3.6), and the manner in which it led to the equations for T_c and Δ, it will be seen that SMW addressed a far more complicated situation. This situation was analysed not via the variational approach followed in BCS theory, but via an elegant alternative approach provided by the Bogoliubov–Valatin transformation.[2]

(iii) Two gaps and, in general, two T_cs arise in this approach because the BCS interaction parameter λ is found to be given by a quadratic equation involving V_{ss}, V_{dd} and V_{sd}, which denote the interaction energies between pairs of electrons in the three pieces of the Hamiltonian.

(iv) SMW have asserted that

(a) When $V_{sd} = 0$, the SC is characterized by two gaps and two T_cs;

(b) When $V_{ss} = V_{dd} = 0$ and $N_s \neq N_d$, there are still two gaps with $\sqrt{N_s N_d}$ as the state density, and

(c) When V_{sd} is finite but much less than $\sqrt{V_{ss} V_{dd}}$, both the gaps close at the same temperature.

The short length of the paper in which the SMW approach was presented is somewhat deceptive: Working through its steps takes some effort! For the purpose of a critical appraisal of the assertions noted above, it suffices to make Eqs. (4) in the SMW paper as our starting point:

$$A(T)[1 - V_{ss}N_sF(A, T)] = B(T)V_{sd}N_dF(B, T),$$
$$B(T)[1 - V_{dd}N_dF(B, T)] = A(T)V_{sd}N_sF(A, T), \tag{5.1}$$

where $A(T)$ denotes the T-dependent gap satisfying

$$1 - \lambda_A F(A, T) = 0, \tag{5.2}$$

$$F(A, T) = \int_0^{k_B \theta_D} d\varepsilon \frac{\tanh(\sqrt{\varepsilon^2 + A^2}/2k_B T)}{\sqrt{\varepsilon^2 + A^2}}, \tag{5.3}$$

and $B(T)$ is defined similarly.

$F(A, T)$ is therefore given by the inverse of the coupling constant. This equation may be used to determine λ_A at any T because λ is independent of T in the BCS theory. In particular, if T_{c1} is the critical temperature for gap A, then $A(T_{c1}) = 0$ and we have $1 - \lambda_A F(0, T_{c1}) = 0$. Similarly, for gap B we have $1 - \lambda_B F(0, T_{c2}) = 0$. Note that $F(0, T_{c1}) \neq F(0, T_{c2})$.

Consider now the case when $V_{sd} = 0$. In this case we have from Eqs. (5.1):

$$A(T)[1 - V_{ss}N_s F(A, T)] = 0,$$
$$B(T)[1 - V_{dd}N_d F(B, T)] = 0.$$

In general therefore, when $A(T) \neq B(T) \neq 0$, we have

$$[1 - V_{ss}N_s F(A, T)] = 0,$$
$$[1 - V_{dd}N_d F(B, T)] = 0,$$

and, in particular

$$[1 - V_{ss}N_s F(0, T_{c1})] = 0,$$
$$[1 - V_{dd}N_d F(0, T_{c2})] = 0.$$

These equations are independent of each other and lead to two unequal gaps and the associated T_cs. Hence SMW assertion iv (a) is seen to be valid.

Let us now consider Eqs. (5.1) when $V_{ss} = V_{dd} = 0$. Multiplying the two equations, we then have

$$F(A, T)F(B, T) = 1/V_{sd}^2 N_s N_d \equiv 1/\lambda_{\text{eff}}^2 \quad \text{or,}$$
$$\lambda_{\text{eff}} = +V_{sd}\sqrt{N_s N_d},$$

since λ_{eff} must be positive. So, in this case there is only one gap — not two as stated in iv (b) above.

The values of the gaps corresponding to $V_{sd} = 0$ were designated as A and B. We now consider the case when $V_{sd} \neq 0$; let gap-values now be

A', B'. From Eqs. (5.1) we then have

$$A'(T)[1 - V_{ss}N_sF(A', T)] = B'(T)V_{sd}N_dF(B', T),$$
$$B'(T)[1 - V_{dd}N_dF(B', T)] = A'(T)V_{sd}N_sF(A', T).$$

With the definitions

$$\lambda_s \equiv V_{ss}N_s, \quad \lambda_d \equiv V_{dd}N_d, \quad \alpha \equiv V_{sd}^2/V_{ss}V_{dd},$$
$$F(A', T) \equiv 1/\lambda_1, \quad F(B', T) \equiv 1/\lambda_2, \tag{5.4}$$

these may be written as

$$A'(T)(1 - \lambda_s/\lambda_1) = B'(T)V_{sd}N_d/\lambda_2,$$
$$B'(T)(1 - \lambda_d/\lambda_2) = A'(T)V_{sd}N_s/\lambda_1.$$

Multiplying these together, we obtain

$$(1 - \lambda_s/\lambda_1)(1 - \lambda_d/\lambda_2) = \alpha\lambda_s\lambda_d/\lambda_1\lambda_2, \tag{5.5}$$

where we have used the definitions given in Eq. (5.4).

Note that for any assigned value of α, the single Eq. (5.5) cannot determine the two unknowns λ_1 and λ_2; also that one may not use $\lambda_2 = \lambda_d$ or $\lambda_1 = \lambda_s$ as a first approximation because it causes the LHS of the equation to vanish.

If we now assume that

$$F(A' = 0, T_{c1}) = 1/\lambda_1 = F(B' = 0, T_{c2}) = 1/\lambda_2 = 1/\lambda, \tag{5.6}$$

and recall from above that we had two distinct gaps in the absence of the inter-band interaction, then Eq. (5.6) is *tantamount to demanding* that these gaps close at the same T_c when such an interaction is switched on. Equation (5.5) now becomes

$$\lambda^2 - (\lambda_s + \lambda_d)\lambda + (1 - \alpha)\lambda_s\lambda_d = 0, \tag{5.7}$$

solutions of which are:

$$\lambda_{1,2} = \frac{\lambda_s + \lambda_d \pm \sqrt{(\lambda_s - \lambda_d)^2 + 4\alpha\lambda_s\lambda_d}}{2}. \tag{5.8}$$

In the SMW paper, the equation for F(0) is

$$[V_{ss} + N_d(V_{sd}^2 - V_{ss}V_{dd})F(0)][V_{dd} + N_s(V_{sd}^2 - V_{ss}V_{dd})F(0)] = V_{sd}^2, \tag{5.9}$$

the solutions of which are

$$F(0) = \frac{1}{\lambda_{1,2}} = \left\{ \frac{\pm[V_{sd}^2/N_s N_d + \frac{1}{4}(V_{dd}/N_s - V_{ss}/N_d)^2]^{1/2}}{V_{sd}^2 - V_{ss}V_{dd}} \right.$$

$$\left. - \frac{\frac{1}{2}(V_{dd}/N_s + V_{ss}/N_d)}{V_{sd}^2 - V_{ss}V_{dd}} \right\}. \tag{5.10}$$

This expression has been used by SMW in the BCS equation for T_c — without the $\pm$ signs, which seems to be an inadvertent omission. It is straightforward to check by rationalizing the expression for $1/F(0)$ that, in terms of λ_s, λ_d and α defined in Eq. (5.4), the solutions in Eq. (5.10) are *identical* with those given in Eq. (5.8). It therefore follows that in obtaining Eq. (5.10) SMW have made the assumption stated in Eq. (5.6).

5.3. Application of SMW Approach to MgB$_2$

We now attempt to apply the SMW formalism to the concrete example of MgB$_2$ the two gaps of which have been reported[3] to close at the same T_c. Relevant superconducting features of MgB$_2$ given in the previous chapter are:

$$\Delta_{01} = 2.1 \text{ meV}, \quad \Delta_{02} = 6.2 \text{ meV}; \quad T_c = 39 \text{ K}; \quad \theta_D = 815 \text{ K}. \tag{5.11}$$

In view of assertion iv (a) above, we can calculate the λ_s corresponding to these Δs via

$$\lambda(0) = \frac{1}{\text{arcsin} \, h(k_B \theta_D / \Delta_0)}. \tag{5.12}$$

Thus,

$$\lambda_1 = \lambda_d = 0.238, \quad \lambda_2 = \lambda_s = 0.32. \tag{5.13}$$

We note that the $T_c s$ corresponding to these λ_s are 13.9 and 40.9 K, respectively. On the other hand, λ corresponding to the empirical value of $T_c = 39$ K calculated via

$$\lambda(T_c) = \frac{-1}{\ln(T_c/1.14\Theta_D)} \tag{5.14}$$

is found to be

$$\lambda(39\text{ K}) = 0.315. \tag{5.15}$$

Comparing this value with the second of the two values in Eq. (5.13), it is seen that $\lambda(0)|_{\Delta 2} > \lambda$ (39 K). This inequality will be further discussed below.

We now seek to find if a small positive value of α (which we recall denotes $V_{sd}^2/V_{ss}V_{dd}$) can bring about closure of the two gaps at the higher T_c in accord with the SMW assertion iv (c). To this end we successively put $\alpha = 0.01, 0.02$ and 0.03 in Eq. (5.8) and determine the values of $\lambda_{1,2}$ that they lead to, and find that

$$\begin{aligned} \lambda_{1,2} &= 0.230, \quad 0.328 \ (\alpha = 0.01), \\ \lambda_{1,2} &= 0.222, \quad 0.336 \ (\alpha = 0.02), \\ \lambda_{1,2} &= 0.216, \quad 0.342 \ (\alpha = 0.03). \end{aligned} \tag{5.16}$$

As a matter of fact, smaller the value of α, closer are the $\lambda_{1,2}$-values to the original values in Eq. (5.13) that one had with $\alpha = 0$. It is easy to see that the following inferences drawn from the above considerations are not dependent on the particular values of λ_s invoked in Eq. (5.13): (a) If experiment dictates that an SC has two zero-T gaps but only one T_c, then we need two λs at $T = 0$ which must converge to the same value as T_c is approached. This requirement cannot be met by the SMW approach because any real, finite (non-zero), positive value of α in this approach causes the *greater* of the two λ_s in the theory to *increase* and the smaller one to decrease: This is transparent from (8) above which, as already noted, is identical with (10). (b) *Even if one assumes that a suitable value of α due to the inter-band interactions could meet the requirement under consideration, the fact would remain that a λ at $T = 0$ goes over to another value as T_c is approached.* So either λ_s or λ_d, or both of them must have a rather complicated T-dependence (caused by itinerancy?) if one is to realize the broken curve in Fig. 2 of SMW. (c) The assertion about the nature of T-dependence of the λ_s follows from the fact the figure under consideration shows a change in the curvature of the plot of $\Delta(T)/\Delta_0$ against T near the higher T_c. Finally, (d) if one follows this plot backwards to $T = 0$, one finds a *third* value for Δ_0.

Despite some ambiguous features noted above, SMW approach provides an important clue for dealing with a superconductor that is characterized by two $\Delta_0 s$ that close at the same T_c: consider the possibility of the λ_s in the theory to be T-dependent. This is the feature of their approach that seems to have escaped serious attention and which we follow up below.

5.4. A Quantitative Study of MgB$_2$ via GBCSEs

It was shown in Appendix 3A that the value of λ for any SC calculated via the equation for Δ_0 is invariably different from its value calculated via the equation for T_c. This implies that λ must be temperature-dependent — a feature we also inferred from the SMW approach if the two gaps of an SC close at the same T_c. We now revisit the superconductivity of MgB$_2$ in the light of this feature and write the equations for its two gaps as ($|W_1| = \Delta_1$, $|W_2| = \Delta_2$):

$$1 = \lambda_1^c(T) \int_{|W_1|/2}^{k_B\theta_1^c + |W_1|/2} dx \frac{\tanh(x/2k_BT)}{x}, \tag{5.17}$$

$$1 = \lambda_1^c(T) \int_{|W_2|/2}^{k_B\theta_1^c + |W_2|/2} dx \frac{\tanh(x/2k_BT)}{x}$$

$$+ \lambda_2^c(T) \int_{|W_2|/2}^{k_B\theta_2^c + |W_2|/2} dx \frac{\tanh(x/2k_BT)}{x}. \tag{5.18}$$

The equation for T_c following from Eq. (5.18) by putting $|W_2| = 0$ is:

$$1 = \lambda_1^c(T_c) \int_0^{k_B\theta_1^c} dx \frac{\tanh(x/2k_BT_c)}{x} + \lambda_2^c(T_c) \int_0^{k_B\theta_2^c} dx \frac{\tanh(x/2k_BT_c)}{x}. \tag{5.19}$$

For convenience, we drop the superscripts c of all the symbols from now on. The values of the Debye temperatures in the above equations were obtained in the last chapter as:

$$\theta_1 = 1062 \text{ K}, \quad \theta_2 = 322 \text{ K}.$$

The assumption that the λ_s above are independent of T then led, via solution of the simultaneous Eqs. (5.18) and (5.19) with the input of $|W_{20}| = 6.28$

meV into the former at $T = 0$ and $T_c = 39$ K into the latter, to the following values:

$$\lambda_1 = 0.2216, \quad \lambda_2 = 0.1073. \tag{5.20}$$

We also recall that with (θ_1, λ_1)-values as $(1062 \text{ K}, 0.2216)$, solution of Eq. (5.17) at $T = 0$ and that of Eq. (5.19) with *only the first term* led to the following results:

$$|W_{10}| = \Delta_{10} = 2.03 \text{ meV}, \quad T_{c1} = 13.2 \text{ K}. \tag{5.21}$$

Above results imply that Δ_{20} closes at $T = 39$ K and Δ_{10} at $T = 13.2$ K. If experiment dictates that both the gaps close at $T = 39$ K, then — guided by the clue provided by the SMW approach — we need to the make the $\lambda_s T$-dependent as below:

$$\lambda_1^c(T) = \lambda_1^c(0) + \alpha_1 T, \quad \lambda_1^c(0) = 0.2216, \quad \alpha_1 = 1.7923 \times 10^{-3} \text{ K}^{-1},$$
$$\lambda_2^c(T) = \lambda_2^c(0) + \alpha_2 T, \quad \lambda_2^c(0) = 0.1073, \quad \alpha_2 = -2.749 \times 10^{-3} \text{ K}^{-1}. \tag{5.22}$$

Note that the values of $|W_{10}|$ and $|W_{20}|$ are unaffected by the above T-dependence. Plots of $w_1(t) = |W_1(T)|/|W_{20}|$ and $w_2(t) = |W_2(T)|/|W_{20}|$ against $t = T/T_c$, with $W_{20} = 6.28$ meV and $T_c = 39$ K, have been given in Fig. 5.1.

5.5.   Remarks

1. It was noted above that the value of λ for any SC calculated via the equation for Δ_0 is invariably different from its value calculated via the equation for T_c This is so even after the error bars on the experimental values of the parameters are taken into account as is evidenced by the values for the "bad actor" SCs, such as Pb and Hg. Since the difference between the two λ-values for the "good" SCs is small, it may be eliminated from the equations for Δ_0 and T_c, whence one is led to the well-known BCS value of about 3.53 for the gap-to-T_c ratio. However, this cannot be done for the "bad actor" SCs. T-dependence of λ is thus seen to shed light on the violation of the alleged universal gap-to-T_c ratio in BCS theory.

2. There is a feature of the SMW approach that needs to be pointed out: since it is based on the usual BCS equations which, as was shown in

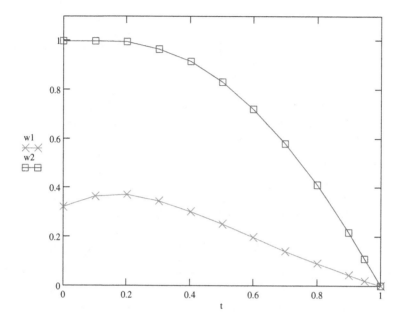

Fig. 5.1. Plots of $w_1(t) = |W_1(T)|/|W_{20}|$ and $w_2(t) = |W_2(T)|/|W_{20}|$ against $t = T/T_c$ obtained via solutions of Eqs. (5.17) and (5.18), respectively ($W_{20} = 6.28$ meV and $T_c = 39$ K).

chapter 3, correspond to CPs bound via OPEM, it cannot account for the observed high-T_cs. For such T_cs, we *need* CPs to be bound together with glue stronger than the one provided by OPEM.

3. It was seen that the SMW approach is characterized by three interaction parameters: V_{ss}, V_{dd}, and V_{sd}, whereas the approach based on GBC-SEs seems to contain only two: λ_1 and λ_2. In this connection we note that while the physical origin of multiple gaps in the two approaches is somewhat different, the latter approach nonetheless has, effectively, three interaction parameters for a 2-gap SC. Symbolically, these are:

$$(\lambda_1, \theta_1), \qquad (5.23)$$

$$(\lambda_2, \theta_2), \qquad (5.24)$$

$$(\lambda_1, \theta_1) + (\lambda_2, \theta_2). \qquad (5.25)$$

While Eq. (5.25) leads to $|W_{20}| = 6.28$ meV and $T_c = 39$ K, $\lambda_1 = 0.2216$ and $\theta_1 = 1062$ K in (5.23) were shown to lead to $|W_{10}| = 2.03$ meV, $T_{c1} = 13.2$ K (see Eq. 5.21). We now note that $\lambda_2 = 0.1073$ and $\theta_2 = 322$

K in Eq. (5.24) lead to

$$|W_{10}| = \Delta_{10} = 4.975 \times 10^{-3} \text{ meV}, \quad T_{c1} = 0.033 \text{ K}. \quad (5.26)$$

4. In our earlier work,[4] we had remarked that the value of the gap in Eq. (5.26) is "too small to be observed." However, precisely such gaps have recently been reported[5] for the iron-pnictide family, a prominent member of which, $Ba_{0.6}K_{0.4}Fe_2As_2$, was dealt with[6] via GBCSEs in the last chapter.

5. The results obtained for MgB_2 via GBCSEs in this chapter will be used for a study of its thermal conductivity in the next chapter.

6. Introduction of T-dependent interaction parameters as in Eq. (5.22), which we were led to via the clue provided by the SMW approach, differs from the usual practice of investigating the effect of temperature on any property of the system by averaging over the Boltzmann-weighted T-independent energy eigenvalues. Nonetheless, it is not the first time that such territory has been reached, as is evidenced by employment of T-dependent dynamics in the context of:

   (a) Superconductivity in the work of Bogoliubov, Zubarev, and Tserkovnikov, as discussed by Blatt.[7]

   (b) An explanation[8] of the empirical law: $H_c(T) = H_c(T = 0)[1 - (T/T_c)^2]$, where $H_c(T)$ is the critical field at T.

   (c) Finite-temperature behaviour[9] of a class of relativistic field theories (RFTs) to address the question of restoration of a symmetry which at zero temperature is broken either dynamically or spontaneously.

   (d) Wick-Cutkosky model[10] in an RFT.

   (e) The legions of unidentified solar emission lines.[11]

   (f) QCD to explain[12] the masses of different quarkonium families and their de-confinement temperatures, and more recently in.

   (g) A comparative study[13] of the experimental features of the Bose–Einstein condensates of ^7Li, ^{23}Na, ^{41}Ca, ^{85}Rb, ^{87}Rb and ^{133}Cs.

7. Finally, we draw attention to a recent paper[14] where electron–phonon coupling in $La_{2-x}Sr_xCuO_4$ probed by ultrafast X-ray diffraction has been reported to be temperature-dependent. In this experimental set up thermally activated phonons could be distinguished from those emitted by

conduction electrons. Thus, separating the electronic from the out-of-equilibrium lattice sub-systems, the authors were able to probe their reequilibration in $La_{2-x}Sr_xCuO_4$ single crystals by monitoring the transient temperature via femtosecond X-ray diffraction for $x = 0.1$ and 0.21. For $x = 0.1$, λ has been reported to vary between 0.0131 and 1.0587.

Notes and References

1. H. Suhl, B.T. Matthias and L.R. Walker, *Phys. Rev. Lett.* **3**, 552 (1959).
2. N.N. Bogoliubov, *Nuovo Cimento* **7**, 794 (1958);
   J.G. Valatin, *Nuovo Cimento* **7**, 843 (1958).
3. M. Iavarone, G. Karapetrov, A.E. Koshelev, W.K. Kwok, G.W. Crabtree and D.G. Hinks, *Phys. Rev. Lett.* **89**, 187002 (2002).
4. G.P. Malik and U. Malik, *J. Supercond. Nov. Mag.* **24**, 255 (2011).
5. D. Lee, *Nature Physics* **8**, 364 (2012);
   Y. Zhang, Z.R. Ye, Q.Q. Ge, F. Chen, J. Jiang, M. Xu, B.P. Xie and D.L. Feng, *Nature Phys.* **8**, 371 (2012);
   M.P. Allan, A.W. Rost, A.P. Mackenzie, Y. Xie, J.C. Davis, K. Kihou, C.H. Lee, A. Iyo, H. Eisaki and T.M. Chuang, *Science* **336**, 563 (2012).
6. G.P. Malik, I. Chávez, and M. de Llano, *J. Mod. Phys.* **4**, 474 (2013).
7. J.M. Blatt, *Theory of Superconductivity* (Academic Press, New York, 1964), p. 250.
8. G.P. Malik, *Physica B* **405**, 3475 (2010).
9. D. Kirzhnits, A.D. Linde, *Phys. Lett. B* **42**, 471 (1972);
   A.D. Linde, *Rep. Prog. Phys.* **42**, 389 (1979);
   S. Weinberg, *Phys. Rev. D* **9**, 3357 (1974); L. Dolan and R. Jackiw, *Phys. Rev. D* **9**, 3320 (1974).
10. G.P. Malik and L.K. Pande, *Phys. Rev. D* **37**, 3742 (1988).
11. Main reference to this body of work is: G.P. Malik, L.K. Pande, and V.S. Varma, *Astrophys. J.* **379**, 795 (1991).
12. G.P. Malik, R.K. Jha and V.S. Varma, *Eur. Phys. J. A* **2**, 105 (1998); **3**, 373 (1998).
13. G.P. Malik and V.S. Varma, *Int. J. Mod. Phys. B* **27**, 1350042 (10 pages) (2013).
14. B. Mansart *et al.*, *Phys. Rev. B* **88**, 054507 (2013).
15. This chapter is based mainly on: G.P. Malik and M. de Llano, *Int. J. Mod. Phys. B* **29** 1347008 (9 pages) (2013).

Chapter 6

Thermal Conductivity of MgB$_2$

6.1. Introduction

The detailed information about the temperature-dependent gaps of MgB$_2$ that we were led to via GBCSEs in the previous two chapters finds an immediate application in a study of its thermal conductivity, to which this chapter is devoted.

Thermal conductivity (κ) is a measure of heat transport in unit time across any cross-section of a material when it is subjected to a temperature-gradient. Therefore κ is a property which is measured under non-equilibrium conditions. Since heat is carried both by electrons and phonons, $\kappa = \kappa_e + \kappa_g$, where κ_e and κ_g denote the electronic and the phonon or lattice parts, respectively. A basic statement about κ is: It becomes finite due only to *irregularities* of the lattice of the material because in a perfect periodic lattice electrons and phonons can move freely. These irregularities, which affect scatterings of both types of heat carriers, can be categorised into: (a) thermal motions of ions and (b) impurities, strains, displaced atoms, etc. Therefore, in principle, κ of any material can be determined by solving a Boltzmann equation for κ_e and another such equation for κ_g by including in these equations collision terms due to all the irregularities. Even for a material in the normal state, it has so far not been possible to do so in practice. However, one can set up and solve the Boltzmann equation corresponding to some special circumstances by assuming that these cause one or the other form of irregularity to play a predominant role.

This chapter is organized as follows. In Sec. 6.2, we first deal with features of the superconducting state that make the calculation of κ an even more complicated problem than for the normal state. This is followed by recalling the expressions (a) for κ_{es}, the electronic part of the thermal conductivity in the superconducting state, first obtained by Geilikman[1] (G) by assuming that electrons are scattered predominantly by impurities, and (b) κ_{gs}, the lattice part of the thermal conductivity in the superconducting state, first obtained by Geilikman and Kresin (GK) by assuming that phonons are scattered predominantly by electrons. While the formidable theory that leads to these expressions is outside the scope of this monograph, we note that a detailed account of it can be found in Bardeen, Rickayzen, and Tewordt[3] (BRT), who independently derived the results obtained by G and GK.

Based on the results obtained for MgB_2 via GBCSEs in the previous two chapters and the G and GK equations, we address in Sec. 6.3 the experimental data on its thermal conductivity obtained by Bauer *et al.*[4] (B *et al.*), and by Schneider *et al.*[5] (S *et al.*). Indeed, employing (G and GK)/BRT equations, both B *et al.* and S *et al.* also carried out similar studies for their respective data. However, these studies were carried out *without* a detailed knowledge of the T-dependent values of the two gaps of MgB_2. Such an input, which is made available by GBCSEs, is *essential* in order to obtain quantitative results for κ_{es} and κ_{gs} and hence for κ_s.

6.2. Features of Superconducting State that Affect Thermal Conductivity

(i) We recall that in the kinetic theory, κ is proportional to the following properties of heat carriers: (a) number density (b) specific heat and (c) mean free path (MFP). Such a relation holds both for the normal and the superconducting states.

(ii) A non-trivial statement about MFP in any system is that it remains unchanged[6,3] when the system passes from the normal to the superconducting state. This is not so for the other two properties on which κ depends.

(iii) The superconducting state is distinguishable from the normal state by the feature of being characterized by a T-dependent gap $\Delta(T)$. For the latter state which exists when $T > T_c$, $\Delta(T) = 0$; as T is decreased below T_c, there is a progressive increase in the value of $\Delta(T)$ till it attains its maximum value at $T = 0$. This increase is synonymous with an increase in the number of Cooper pairs (CPs) and hence a *decrease* in the number density of electrons. Since CPs have zero entropy, they cannot act as carriers of heat. Thus $T_c \geq T \geq 0$ is the region in which both the number density of carriers and their specific heat decrease.

(iv) A consideration of the previous paragraph suggests that there should be a *pronounced* decrease in the value of κ as T approaches 0 K. This usually does not happen because as T is decreased, there is a progressive increase in the value of MFP.

(v) κ is a function of T because both κ_{es} and κ_{gs} depend on T via their dependence on $\Delta(T)$.

(vi) If an SC has more than one gap, then dependence of κ_{es} and κ_{gs} on both these gaps has to be taken into account.

6.3. Equations for κ_{es} and κ_{gs} in the G and GK Theory

The G equation for $\kappa_{es}(T)$ when electrons are scattered predominantly by impurities is[1]:

$$\kappa_{es}(T) = A' F_{es}(T), \qquad (6.1)$$

where A' is independent of T, and

$$F_{es}(T) = k_B T \int_{u(T)}^{\infty} dx \frac{x^2}{(1+e^x)(1+e^{-x})}, \quad (u(T) = \Delta(T)/k_B T). \quad (6.2)$$

The GK equation for κ_{gs} when phonons are scattered predominantly by electrons is[2]:

$$\kappa_{gs}(T) = B' T^2 F_{gs}(T), \qquad (6.3)$$

where B' is independent of T, and

$$F_{gs}(T) = \int_{0}^{2u(T)} dx x^4 e^x \Big/ \left[(e^x - 1)^2 \left\{ 2x - 2\ln\left(\frac{e^{x+u(T)}+1}{e^{x-u(T)}+1}\right) \right\} \right]$$

$$+ \int_{2u(T)}^{\infty} dx x^4 e^x \Big/ \left[(e^x - 1)^2 \left\{ x + 2u(T) - 2\ln\left(\frac{e^{x+u(T)}+1}{e^{x-u(T)}+1}\right) \right\} \right].$$

$$(6.4)$$

It is seen from the above expressions that knowledge of $\Delta(T)$ is *essential* for the determination of $\kappa_{es}(T)$ and $\kappa_{es}(T)$. We now take up the application of above equations to MgB$_2$.

6.4.   On the Input of $\Delta(T)$ of MgB$_2$ into the Equations for κ_{es} and κ_{gs} in the G and GK theory

The T-dependence of the two gaps of MgB$_2$ was dealt with in Chapters 4 and 5; it is governed by the following GBCSEs (see Eqs. 5.17 and 5.18):

$$1 = \lambda_1^c(T) \int_{|W_1|/2}^{k_B\theta_1^c+|W_1|/2} dx \frac{\tanh(x/2k_BT)}{x} \tag{6.5}$$

$$1 = \lambda_1^c(T) \int_{|W_2|/2}^{k_B\theta_1^c+|W_2|/2} dx \frac{\tanh(x/2k_BT)}{x}$$

$$+ \lambda_2^c(T) \int_{|W_2|/2}^{k_B\theta_2^c+|W_2|/2} dx \frac{\tanh(x/2k_BT)}{x}. \tag{6.6}$$

In these equations $|W_1|$ and $|W_2|$ are to be identified with Δ_1 and Δ_2, respectively,

$$\theta_1^c = 1062 \text{ K}, \quad \theta_2^c = 322 \text{ K}, \tag{6.7}$$

and

$$\lambda_1^c(T) = \lambda_1^c(0) + \alpha_1 T, \quad \lambda_1^c(0) = 0.2216, \quad \alpha_1 = 1.7923 \times 10^{-3} \text{ K}^{-1},$$
$$\lambda_2^c(T) = \lambda_2^c(0) + \alpha_2 T, \quad \lambda_2^c(0) = 0.1073, \quad \alpha_2 = -2.749 \times 10^{-3} \text{ K}^{-1}. \tag{6.8}$$

When $T = 0$, the input of Eqs. (6.7) and (6.8) into Eqs. (6.5) and (6.6) yields

$$|W_{10}| = 2.03 \text{ meV}, \quad |W_{20}| = 6.28 \text{ meV}, \tag{6.9}$$

which are in accord with the experimental values. Further, it was shown in chapter 5 that when $T \neq 0$, Eqs. (6.5–6.8) lead to closure of both the gaps at 39 K, which is generally believed to be the case. Therefore, in Scenario 1, we shall use these equations to calculate the values of $\Delta_1(T)$ and $\Delta_2(T)$ which are required as input into Eqs. (6.1) and (6.3).

Keeping an open mind about closure of both the gaps at the same T_c, we also consider here the possibility that λs above are independent of T — as was done in Chapter 4. In this case, which we refer to as Scenario 2, the smaller gap closes at 13.2 K and the larger gap at 39 K. This issue will be further discussed below.

6.5. Addressing the Thermal Conductivity Data on MgB₂ Obtained by B *et al.*[4] via the G[1] and GK[2] Theory and GBCSEs

B *et al.*[4] obtained values of the total thermal conductivity κ_{ts} of MgB₂ at 38 temperatures between 38.7056 K and 7.8313 K. We address these data as follows.

In Scenario 1:

(i) At each temperature in the data, we calculate $\Delta_1(T)$ via Eq. (6.5) and $\Delta_2(T)$ via Eq. (6.6) by using Eq. (6.7) and the T-dependent values of λs given in Eq. (6.8).

(ii) Corresponding to each pair of $[T, \Delta_1(T)]$ and $[T, \Delta_2(T)]$ values from (i), we calculate $F_{es}(T, \Delta_1)$ by using Eq. (6.2) and $F_{gs}(T, \Delta_2)$ by using Eq. (6.4).

(iii) From Eqs. (6.1) and (6.3), for the first gap we have

$$\kappa_{s1}(T) = \kappa_{es1}(T) + \kappa_{gs1}(T) = A' F_{es1}(T, \Delta_1) + B' F_{gs1}(T, \Delta_1),$$

which we write as

$$\kappa_{s1}^r(T) = \frac{\kappa_{s1}(T)}{\kappa_{s1}(T_c)} = \frac{\kappa_{es1}(T) + \kappa_{gs1}(T)}{\kappa_{s1}(T_c)}$$

$$= \left[\frac{A' F_{es1}(T_c)}{\kappa_{s1}(T_c)}\right] \frac{F_{es1}(T)}{F_{es1}(T_c)} + \left[\frac{B' T_c^2 F_{gs1}(T_c)}{\kappa_{s1}(T_c)}\right] \frac{F_{gs1}(T)}{F_{gs1}(T_c)} t^2$$

$$\equiv A \, f_{es1}(t) + B \, t^2 \, f_{gs1}(t), \tag{6.10}$$

where the superscript r on κ_{s1} denotes *reduced* (or normalized) thermal conductivity, $t = T/T_c$, A and B are constants, and f_{es1} and f_{gs1} are dimensionless.

Similarly, for the second gap we have

$$\kappa^r_{s2}(T) \equiv A \, f_{es2}(t) + Bt^2 f_{gs2}(t). \tag{6.11}$$

From Eqs. (6.10) and (6.11), we obtain

$$\kappa^r_s(t) = [\kappa^r_{es1}(t, \, \Delta_1) + \kappa^r_{gs1}(t, \, \Delta_1)] + [\kappa^r_{es2}(t, \, \Delta_2) + \kappa^r_{gs2}(t, \, \Delta_2)]$$

$$= A \, [f_{es1}(t, \, \Delta_1) + f_{gs1}(t, \, \Delta_1)]$$

$$+ Bt^2 [f_{es2}(t, \, \Delta_2) + f_{gs2}(t, \, \Delta_2)]. \tag{6.12}$$

(iv) The values of T and $t = T/T_c$ ($T_c = 39\,$K) have been given in the first two columns of Table 6.1, followed by values of $\Delta_1(T)$, $F_{es1}(T, \, \Delta_1)$, $F_{gs1}(T, \, \Delta_1)$ in the next three columns. Subsequent three columns give the values of $\Delta_2(T)$, $F_{es2}(T, \, \Delta_2)$, and $F_{gs2}(T, \, \Delta_2)$.

(v) Values of the parameters given in the remaining columns in the Table 6.1 are obtained after fixing the values of A and B in the above equations. This is done by appealing to the experimental values of $\kappa_s(T)$ at two temperatures. Let the temperatures chosen for this purpose be 7.8313 and 33.5426 K (each of these is marked with an asterisk in Table 6.1). We then have $t = 0.2008$ at the lower temperature and $t = 0.8601$ at the higher temperature. The experimental values of the total thermal conductivity at these temperatures (given in the last column of the table) are 13.9374 and 127.799 (in appropriate units), respectively. Reduced (or normalized) values of these are obtained by dividing these by the experimental value of the thermal conductivity at $T = T_c$, i.e., 146. 5531. The values of the LHS of Eq. (6.12) are therefore fixed at these temperatures. For each of these temperatures, the values of $f_{es1}(t, \, \Delta_1)$ and $f_{es2}(t, \, \Delta_2)$ on the RHS of Eq. (6.12) are obtained by dividing the corresponding $F_{es1}(T, \, \Delta_1)$ and $F_{es2}(T, \, \Delta_2)$ by $F_{es}(T_c, \, \Delta_1)$, i.e., 5.528 meV. Similarly, the values of $f_{gs1}(t, \, \Delta_1)$ and $f_{gs2}(t, \, \Delta_2)$ on the RHS of Eq. (6.12) are obtained by dividing the corresponding $F_{gs1}(T, \, \Delta_1)$ and $F_{gs2}(T, \, \Delta_2)$ by $F_{gs}(T_c, \, \Delta_1)$, i.e., 7.212. The two equations so obtained can now be solved for A and B. For the temperatures under consideration, we obtain

$$A = 0.50903, \quad B = 7.14796 \times 10^{-5}. \quad \text{(Scenario 1)} \tag{6.13}$$

(vi) Since A and B are now fixed, we can obtain at each of the remaining 36 temperatures for each of the gaps the reduced value of the electronic

Table 6.1. Values of $\Delta_1(T)$, $F_{es}(T, \Delta_1)$, $F_{gs}(T, \Delta_1)$ and $\Delta_2(T)$, $F_{es}(T, \Delta_2)$ and $F_{gs}(T, \Delta_2)$ for all $T \le T_c$ in the data of B *et al.*[4] in: (a) Scenario 1: both the gaps close at 39 K and (b) Scenario 2: Δ_1 closes at 13.2 K while Δ_2 closes at 39 K. Entries corresponding to (b) are given in the lower of the two rows against each T. In both scenarios θs are as given in Eq. (6.7). In the former scenario, $\Delta_1(T)$ and $\Delta_2(T)$ are calculated with T-dependent λs given in Eq. (6.8); in the latter scenario, $\Delta_1(T)$ and $\Delta_2(T)$ are calculated by ignoring the T-dependence of λs. $F_{es}(T)$ and $F_{gs}(T)$ corresponding to each $\Delta(T)$ are calculated via Eqs. (6.1) and (6.13), respectively. Entries marked with (*) are used as input to fix A and B.

T	$t = T/39$	$\Delta_1(T)$ meV	$F_{es}(T, \Delta_1)$ $\times 10^{-3}$	$F_{gs}(T, \Delta_1)$	$\Delta_2(T)$ meV	$F_{es}(T, \Delta_2)$ $\times 10^{-4}$	$F_{gs}(T, \Delta_2)$	$\kappa_{es}(T, \Delta_1)$	$K_{gs}(T, \Delta_1)$ $\times 10^{-3}$	$\kappa_{es}(T, \Delta_2)$	$K_{gs}(T, \Delta_2)$	$\kappa_s(T)|_{th}$	$\kappa_s(T)|_{exp}$
7.8313	0.2008	2.3368	0.4322	381.53	6.212	0.0713	$1.363.10^5$	5.836	0.022	0.096	7.983	13.937	13.9374*
		1.479	0.7859	78.84	6.2583	0.06749	$1.460.10^5$	10.637	1.728	0.091	3.200	13.937	
9.4315	0.2418	2.288	0.7261	176.36	6.155	0.3112	$2.664.10^4$	9.805	0.015	0.42	2.263	12.503	19.2367
		1.1298	1.1416	35.424	6.2206	0.2924	$2.889.10^4$	15.461	1.126	0.396	0.9184	16.777	
10.4677	0.2684	2.2423	0.9342	116.11	6.101	0.6377	$1.185.10^4$	12.615	0.012	0.861	1.24	14.728	24.047
		0.8616	1.4264	15.983	6.1827	0.5961	$1.297.10^4$	19.306	0.626	0.807	0.508	20.635	
12.4586	0.3195	2.1299	1.346	59.65	5.958	1.8266	$3.490.10^3$	18.136	0.0097	2.461	0.567	21.174	33.8843
		0.2573	1.7648	8.404	6.0763	1.6911	3909.9	23.902	0.4661	2.29	0.2164	26.409	
14.4752	0.3712	1.991	1.752	35.38	5.759	3.9804	$1.339.10^3$	23.606	0.0078	5.363	0.294	29.271	41.8619
		0	2.0519	7.212	5.9202	3.6596	1531.6	27.790	0.540	4.956	0.1147	32.862	
15.6230	0.4006	1.9038	1.974	27.85	5.622	5.6777	839.76	26.598	0.0071	7.65	0.215	34.470	48.5756
		0	2.2146	7.212	5.809	5.2084	973.94	29.994	0.629	7.054	0.0849	37.133	
17.4586	0.4477	1.7553	2.311	20.42	5.368	9.0843	430.26	31.138	0.0065	12.24	0.137	43.522	53.7372
		0	2.4748	7.212	5.5972	8.3292	510.51	33.518	0.7855	11.281	0.0556	44.855	
19.4661	0.4991	1.5844	2.656	15.79	5.045	13.651	223.19	35.787	0.0063	18.393	0.089	54.275	67.0045
		0	2.7593	7.212	5.3175	12.566	271.27	37.371	0.9766	17.019	0.0367	54.428	
20.4718	0.5249	1.4966	2.821	14.25	4.865	16.19	164.19	38.01	0.0063	21.814	0.072	59.903	71.5823
		0	2.9019	7.212	5.1586	14.957	201.71	39.302	1.08	20.257	0.0302	59.591	
21.5887	0.5536	1.3982	3.000	12.91	4.654	19.14	118.48	40.422	0.0063	25.789	0.058	66.275	78.5762
		0	3.0602	7.212	4.9675	17.774	147.02	41.446	1.201	24.072	0.0245	65.5443	

(*Continued*)

Table 6.1.   (Continued)

T	$t = T/39$	$\Delta_1(T)$ meV	$F_{es}(T,\Delta_1)$ $\times 10^{-3}$	$F_{gs}(T,\Delta_1)$	$\Delta_2(T)$ meV	$F_{es}(T,\Delta_2)$ $\times 10^{-4}$	$F_{gs}(T,\Delta_2)$	$\kappa_{es}(T,\Delta_1)$	$K_{gs}(T,\Delta_1)$ $\times 10^{-3}$	$\kappa_{es}(T,\Delta_2)$	$K_{gs}(T,\Delta_2)$	$\kappa_s(T)|_{th}$	$\kappa_s(T)|_{exp}$
22.4880	0.5766	1.3186	3.14	12.05	4.474	21.577	92.094	42.308	0.0064	29.073	0.049	71.436	80.5136
		0	3.1877	7.212	4.8044	20.108	115.28	43.153	1.301	27.234	0.0208	70.408	
24.4904	0.628	1.1412	3.445	10.62	4.045	27.037	54.487	46.418	0.0067	36.43	0.034	82.888	91.0589
		0	3.4715	7.212	4.3966	25.55	68.33	47.017	1.546	34.604	0.0146	81.637	
26.5235	0.6801	0.9629	3.746	9.62	3.573	32.379	33.86	50.474	0.0071	43.627	0.025	94.133	98.2931
		0	3.7597	7.212	3.9404	31.0313	41.99	50.92	1.813	42.027	0.0106	92.9595	
26.5235	0.6801	0.9629	3.746	9.62	3.573	32.379	33.86	50.474	0.0071	43.627	0.025	94.133	98.2931
		0	3.7597	7.212	3.9404	31.0313	41.99	50.92	1.813	42.027	0.0106	92.9595	
28.5546	0.7322	0.7884	4.041	8.9	3.066	37.289	22.494	54.448	0.0076	50.243	0.019	104.72	107.6549
		0	4.0476	7.212	3.4314	36.228	30.04	54.819	2.101	49.066	0.009	103.896	
30.5499	0.7833	0.622	4.328	8.38	2.536	41.589	16.125	58.315	0.0082	56.037	0.016	114.38	115.6430
		0	4.3304	7.212	2.8822	40.878	18.74	58.649	2.405	55.364	0.00624	114.022	
32.5361	0.8343	0.462	4.611	7.99	1.98	45.36	12.353	62.129	0.0089	62.118	0.014	123.27	120.3485
		0	4.612	7.212	2.2868	44.969	13.77	62.463	2.728	60.904	0.0052	123.375	
33.5426	0.8601	0.384	4.754	7.82	1.688	47.09	11.033	64.055	0.0092	63.449	0.013	127.799	127.799*
		0	4.7547	7.212	1.9664	46.807	12.05	64.396	2.9	63.394	0.0048	127.799	
34.5088	0.8848	0.310	4.891	7.68	1.402	48.67	10.019	65.901	0.0096	65.578	0.012	131.50	131.9507
		0	4.8916	7.212	1.6468	48.514	10.74	66.25	3.0689	65.705	0.0046	131.963	
36.6796	0.9405	0.151	5.199	7.42	0.759	51.96	8.369	70.051	0.010	70.011	0.012	140.53	140.2963
		0	5.1993	7.212	0.8854	51.936	8.65	70.419	3.467	70.342	0.00403	137.130	
38.7056	0.9925	0.0115	5.487	7.23	0.096	54.86	7.335	73.932	0.011	73.918	0.012	147.94	145.2463
		0	5.4865	7.212	0.1195	54.865	7.37	74.309	3.861	74.309	0.00395	148.625	
39	1	0	5.528	7.212	0	55.28	7.212	74.484	0.011	74.7484	0.011	148.99	146.5531

part of κ by using Eq. (6.10) and the lattice part by using Eq. (6.11). The actual values of these components of κ given in Table 6.1 are obtained by multiplying the reduced values by the corresponding factors that were used to reduce them.

In Scenario 2:

The procedure detailed above is repeated by ignoring the T-dependence of λs in Eq. (6.8). As was noted above, the gaps now close at different temperatures. The values of A and B in this case are found to be

$$A = 0.50637, \quad B = 2.65104 \times 10^{-5}. \qquad \text{(Scenario 2)} \qquad (6.14)$$

Calculated values of various parameters for this scenario are included in Table 6.1: these have been given in the lower of the two rows against each T.

Plots of some of the results obtained in both the scenarios are given in Fig. 6.1.

6.6.   The Data Obtained by Schneider *et al.*[5]

A feature of these data is: At around the same temperatures below 39 K, the values of $\kappa_s(T)$ are significantly lower than the values reported by B *et al.* For example, $\kappa_s(38.86986)|_S = 80.4$ whereas $\kappa_s(38.7056)|_B = 145.2463$ mW cm^{-1} deg^{-1}. Further, while B *et al.* reported their values up to $T = 7.4187$ K, S *et al.* did so up to $T = 1.8752$ K. The values of the parameters A and B obtained with the input of experimental values of $\kappa_s(T)$ at $T = 0.452423$ and 25.84772 are now found to be

$$A = 0.48412, \quad B = 1.01782 \times 10^{-6}. \qquad \text{(Scenario 2)} \qquad (6.15)$$

A plot of the experimental values of S *et al.* and the corresponding theoretical values obtained via the above approach is given in Fig. 6.2.

6.7.   Remarks

1. A general perception about thermal conductivity of a material in the superconducting state is that it belongs more to the realm of applications than theory. This is perhaps one of the reasons that none of the standard texts on superconductivity mentioned in chapter 3 deals with it. Note however that even if these texts were to include such a discussion, they

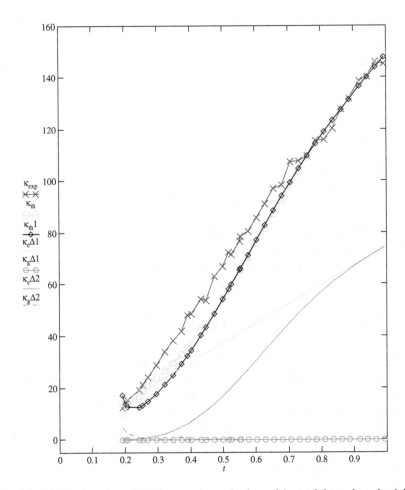

Fig. 6.1.   Clockwise, plots of: (a) the experimental values of the total thermal conductivity κ_S (meVcm^{-1}deg^{-1}) against $t = T/T_c$ (blue) for all $T \leq T_c$ in the data of B *et al.*[4]; (b) the theoretical values of κ_S in Scenario 2 (green); (c) the theoretical values of κ_S in Scenario 1 (black). The remaining four plots correspond to Scenario 2: Uppermost of these is for the electronic part of κ_S due to the smaller gap, followed by the similar part due to the larger gap. The two coincident plots at the bottom of the figure are the lattice parts of the total thermal conductivity due to the two gaps.

would be unable to address the thermal conductivity of composite HTSCs such as MgB$_2$ simply because the usual BCS equations cannot give a *quantitative* account of the observed values of the gaps. We have chosen to include this topic here as an example of how the BSE-based approach

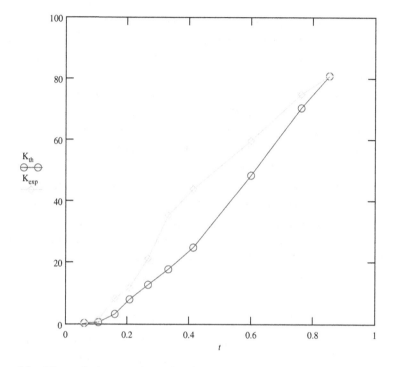

Fig. 6.2. Plots of the experimental values (green) of total thermal conductivity ($meVcm^{-1}deg^{-1}$) in the data of S *et al.*[5] against $t = T/T_c$ at some selected temperatures below T_c and their theoretic al counterparts (blue) in Scenario 2.

which leads to GBCSEs enables one to address problems in a larger arena.

2. A feature of the approach followed above is that we have not relied upon the Wiedemann–Franz law to separate out the lattice part of the total thermal conductivity as is usually done. The applicability of this law to MgB_2 has been questioned.[7]

3. We have given above the values of A and B that occur in Eqs. (6.10–6.12) by appealing to the experimental data at 7.8313 and 33.5426 K. Values obtained by choosing different pairs of temperatures lead to values similar to those given in Eqs. (6.13) and (6.14). Such values are given in the last reference under Notes and References.

4. From Table 6.1 or the plots in Fig. 6.1 it is seen that the major features of the data of B *et al.* are:

(i) Near T_c, in both scenarios, the total thermal conductivity is constituted mostly by the electronic parts and that $\kappa_{es}(T, \Delta_1) \approx \kappa_{es}(T, \Delta_2)$, i.e., both the gaps make nearly equal contributions to the total value of κ_s.

(ii) As T is decreased, in both scenarios, the electronic part of κ_s due to each gap decreases and the lattice part due to each gap increases. However, throughout the range of B's data, κ_s continues to be dominated by the electronic parts. This is understandable because the lowest temperature in the data is 7.8313 K, i.e., still far from 0 K where the electronic parts are expected to become negligible.

(iii) For values of t in the range $1 \geq t \geq 0.45$, the plots for κ_s in the two scenarios are nearly indistinguishable. However, for values of $t < 0.45$, the plot obtained in Scenario 2 is closer to the experimental curve than the plot obtained in Scenario 1. A plausible reason for this follows.

5. It was noted above that thermal conductivity is measured under non-equilibrium conditions, which are not the conditions under which closure of the two gaps at the same T_c has been observed. Since a thermal current tends to drag a small electric current with it, this current must be balanced by an equal and opposite current. In an SC it is balanced by a supercurrent. For these reasons measurement of thermal conductivity requires a rather elaborate experimental set up. It is not inconceivable therefore that the cumulative effect of the stresses caused by this set up lifts the "degeneracy" of the two gaps closing at the same T, which is possibly the cause of Scenario 2 yielding results in better agreement with experiment than Scenario 1.

6. While qualitatively the data of S *et al.* also shows the features noted above for the data of B *et al.*, the two sets of data differ significantly in matters of quantitative detail. This is not surprising because thermal conductivity is a sample-specific property: Each sample is characterized by its own intrinsic irregularities depending upon the manner of its preparation.

7. In their study concerned with the thermal conductivity of single crystalline MgB_2, Sologubenko *et al.*[7] noted that: "Thus, we consider two subsystems of quasiparticles with gaps Δ_1 and Δ_2, different parameters E_1 and E_2 of phonon-electron scattering, and separate contributions κ_{e1} and κ_{e2} to the heat transport." In this manner they were able to give good

estimates of the zero-temperature values of the two gaps. It will thus be seen that while the BSE-based approach given in this chapter employs a similar approach, it gives the qualitative ideas of Sologubenko *et al.* a *tangible* form.

Notes and References

1. B.T. Geilikman, *Sov. Phys. JETP* **7**, 721 (1958).
2. B.T. Geilikman and V.Z. Kresin, *Sov. Phys. Dokl.* **3**, 1161 (1959).
3. J. Bardeen, G. Rickayzen and L. Tewordt, *Phys. Rev.* **113**, 982 (1959).
4. E. Bauer, Ch. Paul, St. Berger, S. Majumdar and H. Michor, *J. Phys.: Condens. Matter* **13**, L 487 (2001).
5. M. Schneider, D. Lipp, A. Gladun, P. Zahn, A. Handstein, G. Fuchs, S-L Drechsler, M. Richter, K.-H. Müller and H. Rosner, *Physica C* **363**, 6 (2001).
6. P. Wyder,*Phys. Konden. Materie* **3**, 292 (1965).
7. A.V. Sologubenko, J. Jun, S.M. Kazakov, J. Karpinski, H.R. Ott, *Phys. Rev. B***66**, 014504 (2002).
8. This chapter is based on: G.P. Malik and Usha Malik, *WJCMP* **4**, 39 (2014).

Chapter 7

Dynamical Equations
for Temperature-Dependent
Critical Magnetic Fields

7.1. Introduction

Critical magnetic field of an elemental SC in BCS theory is usually calculated via the law of corresponding states: If any two parameters from the set $\{T_c, H_0, \gamma\}$ are known, the third can be predicted; these parameters denote, respectively, the critical temperature, critical magnetic field at $T = 0$, and the electronic specific heat constant. The law, obtained via Eq. (3.16) in Chapter 3, is given by the following equation[1]:

$$H_0 = \sqrt{\frac{\gamma}{0.170}} T_c, \tag{7.1}$$

where H_0 and γ are measured in units of gauss and erg/cm^{3}deg^2, respectively. With H_0 known, the critical magnetic field $H_c(T)$ can be estimated by invoking the empirical relation

$$H_c(T) = H_0[1 - (T/T_c)^2]. \tag{7.2}$$

It has been noted[1] that Eq. (7.1) yields values of H_0 for various simple superconductors that are within 10% of the experimental values, and that[2] Eq. (7.2) is "not in general exactly true, but it provides a convenient representation accurate for many purposes."

It seems obvious that one ought not to apply Eq. (7.1) to SCs that are characterized by multiple gaps. Nonetheless let us naively apply it to MgB_2. For this SC, with $\gamma = 0.89$ mJ/gat K^2 (1248 erg/cm^{3}K^2) and $T_c = 36.7$ K, we obtain $H_0 = 3144$ G. This value is not even close to any of the three values that are found via experiment and given in Ref. 3, which also gives the values of γ and T_c used here. Hence there is a need to obtain alternatives to Eq. (7.2) — for elemental as well as non-elemental SCs. We show in this chapter how this need is met via GBCSEs. The equations so obtained may be regarded as another step towards the goal of deriving equations which may be used to unravel the full complex phase diagram of HTSCs.

This chapter is organized as follows. In the next section we obtain for the Δ of an elemental SC an equation that incorporates both T and H. This requires appealing to the Landau quantization (LQ) scheme, besides the Matsubara prescription. The equation so obtained is solved in Sec. 7.3 with the input of H_0 of some select elements. We are thus enabled to compare the parameters characterizing the superconducting states of these elements at $T = 0, H = H_0$ and $T = T_c$ and H = 0. In Sec. 7.4, we solve the equation when T $\neq 0$ and H $\neq 0$ by assuming Eq. (7.2). We find that the empirical law Eq. (7.2) is satisfied only if V (as in [N(0)V]) depends linearly on T. In Sec. 7.5, we obtain the T $\neq 0$, H $\neq 0$ equations for a composite SC. Section 7.6 is devoted to the rather interesting topic concerned with the possibility of occurrence of superconductivity in a material at values of H exceeding its H_0-value. In the final section we deal briefly with another topic that has sporadically been of some interest: The possibility of the T_c(H) plot of an SC exhibiting de Haas–van Alphen oscillations.

7.2. Incorporation of an External Magnetic Field into the T-dependent Gap Equation for Elemental SCs

BCS equation for $\Delta(T)$ may be regarded as the fundamental equation of the theory because the equation for Δ_0 follows from it by putting $T = 0$, as also the equation for T_c by putting $\Delta = 0$ and $T = T_c$. We now wish to incorporate an external magnetic field H into the equation for $\Delta(T)$. With composite SCs in mind, in the following we do so not for the equation for $\Delta(T)$ but an equation equivalent to it, i.e., the equation for W(T). The equivalence of these equations for elemental SCs was dealt with in detail in Chapter 3. As shown in Chapter 4, the equation for a composite SC is obtained by replacing the one-particle propagator in the T-dependent BSE for an elemental SC by a superpropagator.

Incorporating an external field H at any temperature into the pairing equation is a two-step process: (a) temperature-generalization of the $T = 0$ BSE and (b) generalization of the equation obtained in (a) to include an external magnetic field. We have already carried out the first step in Chapter 3 where it led to Eq. (3.24) reproduced below:

$$1 = \frac{V}{(2\pi)^3} \frac{1}{2} \int_{E_F - k_B\theta_D}^{E_F + k_B\theta_D} d^3p \frac{\tanh[(\beta/2)(p^2/2m - E_F - W/2)]}{p^2/2m - E_F - W/2}. \quad (7.3)$$

This equation determines (half) the binding energy $W(T)$ of a CP at any temperature in zero magnetic field. We now embark on step (b) by invoking an important result proved by Rieck *et al.*[4] and stated as: "the ground-state wave function for a particle with charge 2e moving in a constant magnetic field remains a solution of Eq. (1) (equivalent to our Eq. (7.3)) even in the presence of Landau level quantization." For a magnetic field along the z-axis, Eq. (7.3) can therefore be generalized to include the effects of an external magnetic field via the following replacements.[5]

$$\int dp_x dp_y = 2\pi eH \sum_n, \quad \frac{p_x^2}{2m} + \frac{p_y^2}{2m} = (n + 1/2)\hbar\Omega(H),$$

$$\Omega(H) = \Omega_a H, \quad \Omega_a = \frac{e}{mc}. \quad (7.4)$$

The transverse components of momentum are thus quantized into Landau levels, and we obtain

$$1 = \frac{eHV}{8\pi^2} \int_{-L_m}^{L_n} dp_z$$

$$\times \sum_{n=0}^{n_m} \frac{\tanh[(\beta/2)(p_z^2/2m - E_F - W/2 + (n + 1/2)\hbar\Omega_a H]}{(p_z^2/2m - E_F - W/2 + (n + 1/2)\hbar\Omega_a H}, \quad (7.5)$$

where both L_m and n_m are usually ∞. Because energy of an electron in our problem is constrained to lie in a narrow shell around the Fermi surface, we fix L_m and n_m by appealing to the law of equipartition of energy, and splitting the region where $V \neq 0$ in Eq. (7.5) as:

$$\frac{-2k_B\theta_D}{3} \leq \frac{p_x^2 + p_y^2}{2m} \leq \frac{2k_B\theta_D}{3}, \quad \frac{-k_B\theta_D}{3} \leq \frac{p_z^2}{2m} - E_F \leq \frac{k_B\theta_D}{3}. \quad (7.6)$$

The first of these relations fixes n_m as:

$$n_m(H) = floor\left\{\frac{2k_B\theta_D}{3\hbar\Omega(H)} - 1/2\right\}, \quad (7.7)$$

where *floor* denotes lower of the two integers closest to the number yielded by the parentheses.

Defining $p_z^2/2m - E_F = \xi$, whence $dp_z = md\xi/p_z = md\xi/\sqrt{2m(E_F + \xi)}$ $\approx md\xi/\sqrt{2mE_F}$, and noting that the integrand in Eq. (7.5) is an even function of p_z, we obtain

$$1 = \frac{eHV}{4\pi^2}\sqrt{\frac{m}{2E_F}}\int_0^{k_B\theta_D/3} d\xi \sum_0^{n_m(H)}$$
$$\times \frac{\tanh[(\beta/2)(\xi - W/2 + (n + 1/2)\hbar\Omega_a H]}{(\xi - W/2 + (n + 1/2)\hbar\Omega_a H}. \quad (7.8)$$

This equation determines W at any given values of H and T ($H < H_0$, $T < T_c$), provided that the pre-factor of the integral is known.

Equation (7.8) can be recast in terms of a dimensionless variable x defined as $x = \xi/\hbar\Omega(H)$, whence we obtain

$$1 = \frac{eHV}{4\pi^2}\sqrt{\frac{m}{2E_F}}\int_0^{k_B\theta_D/3\hbar\Omega_a H} dx \sum_0^{n_m(H)}$$
$$\times \frac{\tanh[(\beta\hbar\Omega(H)/2)(x - W/2\hbar\Omega_a H + n + 1/2)]}{x - W/2\hbar\Omega_a H + n + 1/2}. \quad (7.9)$$

The integrand in this equation is manifestly dimensionless — so must of course be the pre-factor. Since we wish to retain the units employed in BCS theory, i.e., gauss for H and eV-cm^3 for V, one way to see that the pre-factor is dimensionless is to employ appropriate conversion factors to write both gauss and cm^3 in terms of eV. On doing so (see Units: cgs and natural at the beginning of the monograph), we have

$$\lambda_m(V, H) \equiv \frac{eHV}{4\pi^2}\sqrt{\frac{m}{2E_F}} = d(E_F)\frac{VH}{4\pi}a_2^3 b_1 b_2$$

$$d(E_F) = \frac{1}{2\pi}\sqrt{\frac{2mc^2}{E_F}}. \tag{7.10}$$

Since V and H are now numbers, λ_m is dimensionless. Note that if we divide $d(E_F)$ by $\hbar c$, then we obtain the one-dimensional density of states in units of $eV^{-1}cm^{-1}$. Of course this will require multiplying V in Eq. (7.10) by the same factor which will then have units of eVcm.

The equation for $H_c(T)$, or $T_c(H)$, following from Eq. (7.9) by putting W = 0 and using Eq. (7.10), is:

$$1 = \lambda_m(V, H)\int_0^{k_B\theta_D/3\hbar\Omega_a H} dx \sum_0^{n_m(H)}$$

$$\times \frac{\tanh[\{\hbar\Omega_a H/2k_B T\}(x + n + 1/2)]}{x + n + 1/2}. \tag{7.11}$$

While Eq. (7.9) determines W (which we recall corresponds to Δ) for any given values of T and H ($T < T_c$, $H < H_0$), Eq. (7.11) determines T_c, or H_c, of an element in the superconducting state when it is subjected to an external magnetic field in a heat bath. The former of these is the usual BCS equation for $\Delta(T)$ containing the additional variable H. Likewise the latter equation for T_c now also contains H. We note that experimental values of Δ dependent on both T and H are seldom available.

In the next section we take up solutions of Eq. (7.11) at $T = 0$ for some select elements with a view to comparing the characteristic features of their superconducting states at $T = T_c$, H = 0 and $T = 0$, H = H_0. This will be followed by a study of the same elements at $T \neq 0$, H $\neq 0$ in the BCS regime ($T < T_c$, H < H_0) via Eq. (7.11).

7.3. Superconducting Features of Some Elements in the $T = T_c$, $H = 0$ and $T = 0$, $H = H_0$ States

BCS equation for T_c of an elemental SC when $H = 0$ is:

$$1 = \lambda \int_0^{\theta_D/2T_c} dx \frac{\tanh(x)}{x}. \tag{7.12}$$

With the input of θ_D and T_c of any element into this equation, we first determine its interaction parameter $\lambda \equiv [N(0)V]$. Following a suggestion originally due to Pines,[6] we then decouple it into $N(0)$ and V by means of

$$N(0) = \frac{3\gamma}{2\pi^2 k_B^2 v}, \tag{7.13}$$

where is v the gm-atomic volume of the element. The results so obtained for some select elements have been given in Table 7.1.

We now deal with Eq. (7.11) which at $T = 0$ reduces to

$$1 = \lambda_m(V_0, H_0) \int_0^{k_B \theta_D/3\hbar\Omega_a H_0} dx \sum_0^{n_m(H_0)} \frac{1}{x + n + 1/2}, \tag{7.14}$$

where V_0 has been used to denote the value of V at $T = 0$ because we wish to allow for the possibility that V may depend on T. If we use the following relation involving Euler's ψ-functions

$$\sum_{k=0}^{m} \frac{1}{u + kv} = \frac{1}{v}[\psi(m + 1 + u/v) - \psi(u/v)], \tag{7.15}$$

then we can put Eq. (7.14) into a more compact form:

$$1 = \lambda_m(V_0, H_0) \int_0^{k_B \theta_D/3\hbar\Omega_a H_0} dx[\psi(n_m(H_0) + x + 3/2) - \psi(x + 1/2)], \tag{7.16}$$

We now go over the steps involved in solving this equation by taking the example of Pb, the values of θ_D, E_F, and H_0 of which have been given in Table 7.1. Noting that $\Omega_a = e/mc = 1.758805 \times 10^7 \text{ sec}^{-1} \text{ gauss}^{-1}$, we have $\Omega_a H_0 = 1.412 \times 10^{10} \text{ sec}^{-1}$. Upon using Eq. (7.7), we obtain $n_m(H_0) = 592$ and the upper limit of the integral in Eq. (7.16) as 296.6. We

Table 7.1. Comparison of various parameters characterizing the superconducting state at $T = T_c$, $H = 0$, and at $T = 0$, $H = H_c$ for some select elements. The values of T_c, θ_D, H_0, and E_F have been taken from Charles P. Poole, Jr., *Handbook of Superconductivity* (Academic Press, London, 2000); those of γ and v from K.A. Gschneidner, *Solid State Phys.* **16**, 275 (1964). $n_m(H_0)$ is given by Eq. (7.7); $r_n = r_0\sqrt{2n_m(H_0)} + 1$ is the radius of the largest Landau orbit, where $r_0 = \sqrt{\hbar c/eH_0}$.

Element	T_c (K)	θ_D (K)	$\lambda(T=T_c, H=0) = [N(0)V]$	γ (mJ gat^{-1} K^{-2})	v (cm^3 gat^{-1})	$N(0)$ (eV^{-1} cm^{-3}) (10^{22})	VeV cm^3 (10^{-23})	H_0 (Gauss)	E_F (eV)	n_m (H_0)	$\lambda_m(T=0, H=H_c) = [N(0)V]_m$ (10^{-4})	V_m (eV cm^3) (10^{-19})	$N(0)_{1-d}$ (eV^{-1} cm^{-1}) (10^6)	r_n (cm) (10^{-4})
Pb	7.2	96	0.3682	3.14	18.27	2.196	1.677	803	9.47	592	17.67	6.87	2.65	3.12
Hg	4.15	88	0.3145	2.2	14.09	1.995	1.577	411	7.13	1062	9.86	6.50	3.05	5.83
Sn	3.72	195	0.2448	1.78	16.30	1.395	1.755	305	10.2	3172	3.30	3.51	2.55	11.70
In	3.41	108	0.2792	1.70	15.76	1.378	2.026	282	8.63	1900	5.51	5.82	2.78	9.42
Tl	2.38	79	0.2756	2.83	17.22	2.1	1.313	178	8.15	2201	4.76	7.74	2.86	12.76
Nb	9.25	276	0.2840	7.66	10.83	9.036	0.314	2060	5.32	664	15.76	1.79	3.54	2.06

now numerically solve Eq. (7.16) to obtain $\lambda_m(V_0, H_0) = 1.767 \times 10^{-3}$. With m taken as the free electron mass, Eq. (7.10) leads to

$$V_0 = \frac{4\pi\lambda_m(V_0, H_0)}{d(E_F)a_2^3 b_1 b_2 H_0} = 6.869 \times 10^{-19}, \qquad (7.17)$$

which is a number in units of $eVcm^3$. These results for Pb and five other elements are included in Table 7.1, which also gives values of the one-dimensional density of states, i.e., $d(E_F)/\hbar c$, for each of the elements and values of the largest Landau orbit given by

$$r_n = r_0\sqrt{2n_m(H_0) + 1}, \qquad (7.18)$$

where $r_0 = \sqrt{\hbar c/eH_0} = \hbar c/\sqrt{b_1 b_2 H_0}$, which has the value 2.566×10^{-4} cm for $H_0 = 1$ gauss.

Comparison of the values of r_n with coherence lengths at $T = H = 0$ reveals that a magnetic field generally causes a slight increase in the size of a CP, which is reminiscent of the Zeeman spreading of energy levels.

7.4. The $T \neq 0$, $H \neq 0$ Equation: The BCS Regime

When $H = 0$, the input of θ_D and Δ_0 into the equation for the latter determines the interaction parameter λ of an elemental SC. This value of λ then enables one to calculate T_c of the element via another BCS equation. It is therefore natural to ask: Can one follow a similar procedure when $H \neq 0$? To elaborate, let us consider the example of Pb for which we found $\lambda_m(V_0, H_0) = 1.767 \times 10^{-3}$ for $H_0 = 803$ gauss. From this value, one can determine the part of $\lambda_m(V_0, H_0)$ which is independent of H_0 via Eq. (7.10). The question then is: Can one employ the H_0-independent part of $\lambda_m(V_0, H_0)$ in Eq. (7.11) to find for Pb the value of its $H_c(T)$ at any T ($T < T_c$, $H < H_0$)? This would be possible only if V in Eq. (7.11) were independent of T. It turns out, unfortunately, that V depends on T and does so in a manner that is different for different elements — as we now proceed to show.

To investigate whether or not V depends on T for any element we invoke the empirical law given in Eq. (7.2) which, while not exactly true, is known to be a good first approximation. We now return to Eq. (7.11) and write it in terms of the reduced (or normalized) variables defined as:

$$t = T/T_c(H = 0), \quad h = H/H_0. \tag{7.19}$$

The law Eq. (7.2) then becomes $h(t) = 1 - t^2$. These reduced variables have the same range for all the elements: $0 \le t \le 1$ and $1 \le h \le 0$. We also introduce λ_{red} and $v(t)$ defined as

$$\lambda_{red} = \frac{\lambda(V, H)}{\lambda_0} = v(t)h(t), \quad v(t) = \frac{V(T)}{V_0}, \tag{7.20}$$

where $\lambda_0 = \lambda_m(V_0, H_0)$.

Upon using Eqs. (7.19) and (7.20), we have Eq. (7.11) as

$$E(v, h, t) \equiv 1 - \lambda_0 v(t)h(t) \int_0^{k_B \theta_D/3\hbar\Omega_a h(t)H_0} dx \sum_0^{n_m(h(t)H_0)}$$

$$\times \frac{\tanh[\{\hbar\Omega_a h(t)H_0\}/2k_B t T_c\}(x + n + 1/2)]}{x + n + 1/2}. \tag{7.21}$$

Treating t as the independent variable, this equation may be solved for $v(t)$ of any element with the input of its $\lambda_0 = \lambda_m(T = 0, H = H_0)$ from Table 7.1 and assuming that $h(t) = 1 - t^2$. The results so obtained for Pb are:

t	0.1	0.2	0.3	0.4	0.5	0.6	0.7	0.8	0.9
$v(t)$	1.02	1.03	1.05	1.07	1.09	1.11	1.13	1.15	1.17

This is approximately a linear relationship:

$$v(t) = 1 + pt, \tag{7.22}$$

with $p = 0.193$. Values of $v(t)$ for the other five elements in Table 7.1 are similarly found to vary linearly with t, the values of p being 0.114, 0.043, 0.074, 0.07, and 0.078 for Hg, Sn, In, Tl, and Nb, respectively. Plots of some of these have been given in Fig. 7.1.

Having determined the values of $v(t)$ for the elements under consideration, we can determine the corresponding values of $\lambda_{red}(t)$ by using Eq. (7.20). These values, together with the associated values for $n_m(H)$

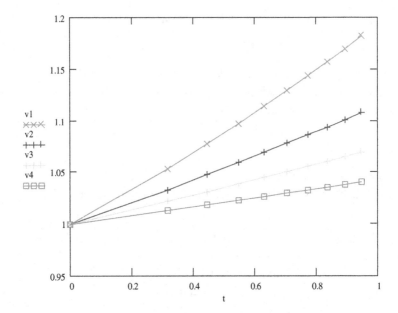

Fig. 7.1.   Plots of $v(t) = V(T)/V_0$ for the values in Table 7.1, which are obtained by solving Eq. (7.21) by assuming $t = \sqrt{1 - h}$. The uppermost curve corresponds to Pb; clockwise, the remaining curves correspond to Hg, In, and Sn.

and $r_n(H)$, have been given in Table 7.2. The plot of $\lambda_{red}(t)$ for Pb is given in Fig. 7.2. It is thus seen that the assumption of the empirical parabolic law Eq. (7.2) has led us to a parabolic form of variation of the magnetic interaction parameter, which is a consequence of a nearly linear variation of V (as in [N(0)V]) with temperature. Since V has turned out to be dependent on T, it is clear that to determine $H_c(T)$ via Eq. (7.11) we cannot employ the H_0-independent part of $\lambda_m(V_0, H_0)$ determined at $T = 0$ via Eq. (7.16).

7.5.   Equations for T-dependent Critical Magnetic Fields of a Composite SC

Assuming that both one- and two-phonon exchange mechanisms may be operative in a composite SC, we consider a sub-lattice of it comprising layers each of which has the structure $A_x B_{1-x}$. While considerations of this section are obviously applicable to MgB_2, we observe that that they can also be applied, for example, to $YBa_2Cu_3O_7$ by identifying Y as A and Ba as

Table 7.2. Variation of various parameters obtained by solving Eq. (7.21) in the BCS regime: $1 \leq h(t) \leq 0$, $t = \sqrt{1-h}$. See Eqs. (7.19) and (7.20) for definitions of t, $h(t)$, $v(t)$, and $\lambda_{red}(t)$ and the caption for Table 7.1 for definitions of $n_m(H)$ and $r_n(H)$.

$h \rightarrow$		1	0.9	0.8	0.7	0.6	0.5	0.4	0.3	0.2	0.1
$t \rightarrow$		0	0.316	0.447	0.548	0.632	0.707	0.775	0.837	0.894	0.949
Element↓ Pb	$n_m(hH_0)$	592	658	741	847	988	1186	1482	1977	2965	5932
	$v(t)$	1	1.0537	1.0777	1.0971	1.1142	1.1294	1.1439	1.571	1.1699	1.1182
	$\lambda_{red}(t)$	1	0.948	0.862	0.768	0.669	0.565	0.438	0.347	0.234	0.118
	$r_n(hH_0)$ (10^{-4}) cm	3.12	3.46	3.9	4.46	5.2	6.24	7.79	10.4	15.6	31.2
Hg	$n_m(hH_0)$	1062	1180	1327	1517	1770	2124	2655	3541	5312	10624
	$v(t)$	1	1.0333	1.048	1.0594	1.0693	1.0782	1.0863	1.0938	1.1009	1.1078
	$\lambda_{red}(t)$	1	0.93	0.838	0.742	0.642	0.539	0.435	0.328	0.22	0.111
	$r_n(hH_0)$ (10^{-4}) cm	5.83	6.48	7.29	8.33	9.72	11.7	14.6	19.4	29.2	58.3
Sn	$n_m(hH_0)$	3172	3524	3965	4531	5287	6344	7931	10575	15863	31726
	$v(t)$	1	1.0133	1.0189	1.0233	1.0269	1.0302	1.0332	1.0359	1.0385	1.041
	$\lambda_{red}(t)$	1	0.912	0.815	0.716	0.616	0.515	0.413	0.311	0.208	0.104
	$r_n(hH_0)$ (10^{-4}) cm	11.7	13.0	14.6	16.7	19.5	23.4	29.3	39.0	58.5	117.0

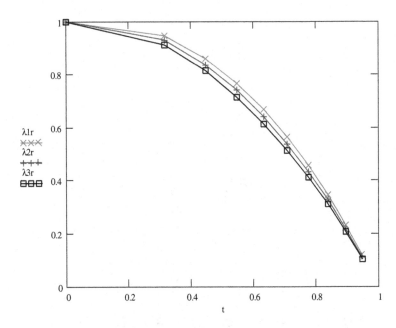

Fig. 7.2.   Plots of $\lambda_{red}(t)$ — defined in (Eq. (7.20) — 0 from the values in Table 7.2 which are obtained by solving Eq. (7.21) by assuming $t = \sqrt{1 - h}$. The uppermost curve corresponds to Pb, followed by the curves for Hg, and Sn.

B, or *vice-versa*, and to $Tl_2Ba_2CaCu_2O_8$ by regarding (Tl, Ca), or (Tl, Ba), or (Ba, Ca) as (A, B). These statements follow from the structures of the unit cells in these SCs, which were dealt with in Chapter 4. The observation made about YBCO and the Tl-based SCs is also applicable to the Bi-based and iron-pnictide SCs.

For pairing via one-phonon exchanges with either the A or B species of ions, we have already obtained Eq. (7.11) for $H_c(T)$ for $T \neq 0$ and Eq. (7.16) for H_0 at $T = 0$. In order to distinguish the situation when A is in its free state from the one in which it not, we now denote the magnetic interaction parameter for the latter by $\lambda_m^{Ac}(V, H)$, just as we did when there was no external magnetic field. Similarly, as in chapter 4, θ_D^A is now denoted by θ_D^{Ac} for A in the composite state. While dealing with pairing via the B ions, A in these symbols is of course replaced by B.

The equation for pairing via simultaneous phonon exchanges with both species of ions is obtained by replacing V in Eq. (7.3) by $V_1 + V_2$. Following then the steps carried out above between Eqs. (7.4) and (7.10), we obtain the required generalization of Eq. (7.11) as

$$
1 = \lambda_m^{Ac}(V, H^{AB}) \int_0^{k_B \theta_D^{Ac}/3\hbar\Omega_a H^{AB}} dx \sum_0^{n_m(H^{AB})}
$$

$$
\times \frac{\tanh[\{\hbar\Omega_a H^{AB}/2k_B T\}(x + n + 1/2)]}{x + n + 1/2}
$$

$$
+ \lambda_m^{Bc}(V, H^{AB}) \int_0^{k_B \theta_D^{Bc}/3\hbar\Omega_a H^{AB}} dx \sum_0^{n_m(H^{AB})}
$$

$$
\times \frac{\tanh[\{\hbar\Omega_a H^{AB}/2k_B T\}(x + n + 1/2)]}{x + n + 1/2}, \tag{7.23}
$$

where $\lambda_m^{Ac(Bc)}$ is the magnetic interaction parameter due to A (B), and H^{AB} is the critical field in the two-phonon exchange scenario. When $T = 0$, Eq. (7.23) reduces to the following equation as a generalization of Eq. (7.16):

$$
1 = \lambda_m^{Ac}(V_0, H_0^{AB}) \int_0^{k_B \theta_D^{Ac}/3\hbar\Omega_a H_0^{AB}}
$$

$$
\times dx[\psi(n_m(H_0^{AB}) + x + 3/2) - \psi(x + 1/2)]
$$

$$
+ \lambda_m^{Bc}(V_0, H_0^{AB}) \int_0^{k_B \theta_D^{Bc}/3\hbar\Omega_a H_0^{AB}}
$$

$$
\times dx \ [\psi(n_m(H_0^{AB}) + x + 3/2) - \psi(x + 1/2)] \tag{7.24}
$$

Note that above considerations have naturally led us to *three* values of the critical field at $T = 0$: two via Eq. (7.16) applied separately to A and B, and one via Eq. (7.24). Note also that H_0 can be separated from $\lambda_m^{Ac \text{ or } Bc}(V_0, H_0)$ in Eq. (7.16), and H_0^{AB} from $\lambda_m^{Ac}(V_0, H_0^{AB})$ and $\lambda_m^{Bc}(V_0, H_0^{AB})$ in Eq. (7.24). In the light of this, for an SC characterized by three values of H_0: $H_{01} < H_{02} < H_{03}$, it is natural to ask:

Q. After determining the H_0-independent parts of $\lambda_m^{Ac \text{ or } Bc}(V_0, H_0)$ with the input of H_{01} or H_{02} into Eq. (7.16), can one employ these to calculate H_{03} via Eq. (7.24)? (7.25)

To answer this question we need to delve on the physical origin of the multiple gaps observed in composite SCs. The equality between Δ and $|W|$

(i.e., half the binding energy of a CP) established in Chapter 3 implies that an SC characterized by, say, three gaps must have CPs that are characterized by three different values for their binding energies. In the binary $A_x B_{1-x}$, this comes about partly because A and B have different θ_Ds and partly because the intrinsic values of V as in [N(0)V] due to A and B are different. In this context it is important to note that the observation of different values of Δ of an SC involves sensitizing the experimental apparatus differently because in doing so one is accessing different parts of the Fermi surface. Examination of the Fermi surfaces[7] of even the elemental SCs reveals that they generally have rather complicated structures. One does not have to bother about them in dealing with elemental SCs because in BCS theory because E_F is not required to solve either the equation for $\Delta(T)$ or the equation for T_c. This is also the case when one deals with composite SCs in the absence of an external magnetic field, as we saw in Chapter 4. To solve the H-dependent equations above, however, E_F is a requisite parameter.

Hence the answer to the question raised in (7.25) is: Yes, *provided* that one can fix the requisite values of E_F — assuming that V can be fixed via the equations for H_0. This issue will be further discussed in the Remarks section.

7.6.   The T $\neq$ 0, H $\neq$ 0 Equation: Beyond the BCS Regime

The possibility of the occurrence of superconductivity in a material at values of H exceeding its H_0-value was noted in Ref. 8; it has also been the subject of investigation in more recent papers.[9] In this section we deal with this aspect of our work, and we do so by returning to Eq. (7.21) without restricting ourselves to the BCS regime defined as $h(t) = 1 - t^2$.

For the sake of concreteness we consider here the case of Sn, and ask about the price one might have to pay in order for it to have a value of $h > 1$. This will require for us to specify the value of t. So, we choose, in the first instance, $t = 0$, which makes the tanh-function in (24) to become unity. Now, fixing h fixes the upper limits of the integral and the sum in this equation. Since the density of states, $d(E_F)/\hbar c$ (see Eq. (7.10)), is independent of both t and h, the only way the equation can be satisfied is via a change in the value of v. This is the price one will have to pay for the desired value of h, and it is calculable. The results have been given in Table 7.3, both for

Table 7.3. Solutions of Eq. (7.21) for v of Sn at $t = 0$ and 0.55 for different values of h, without assuming $h(t) = 1 - t^2$. Also given here are the corresponding numbers of Landau levels, the radii of the smallest and the largest Landau orbits (these do not depend on t), and values of the magnetic interaction parameter, λ_m, defined in Eq. (7.10).

				$t = 0$		$t = 0.55$	
h	n_m ($h\mathrm{H}_0$)	$r_0(h\mathrm{H}_0)$ $\times 10^{-6}$	$r_n(h\mathrm{H}_0)$ $\times 10^{-4}$	$v(h)$	$\lambda_m(h)$ $\times 10^{-3}$	$v(h)$	$\lambda_m(h)$ $\times 10^{-3}$
1	3172	14.69	11.7	1.0000	0.3014	1.0233	0.338
2	1585	10.39	5.85	1.0003	0.6605	1.0234	0.676
3	1057	8.48	3.90	1.0001	0.9905	1.0231	1.013
4	792	7.35	2.92	1.0005	1.321	1.0234	1.352
5	634	6.57	2.34	1.0002	1.651	1.023	1.689
6	528	6.00	1.95	1.0005	1.982	1.0231	2.027
7	452	5.55	1.67	1.001	2.313	1.0236	2.365
8	396	5.19	1.46	1.0004	2.642	1.0229	2.702
9	352	4.90	1.13	1.0004	2.972	1.0227	3.039
10	316	4.65	1.17	1.0014	3.306	1.0237	3.380
50	62	2.08	0.23	1.0088	16.65	1.0269	16.95
100	31	1.47	0.12	1.0077	33.27	1.0209	33.70
500	5	0.66	0.022	1.0848	179.1	1.0853	179.2
1000	2	0.46	0.014	1.1467	378.6	1.1467	378.6
1500	1	0.38	0.0066	1.2093	598.9	1.2093	598.9

$t = 0$, and $t = 0.55$, for a wide range of values of h. Also given there are the corresponding numbers of Landau levels, the radii of the smallest and the largest Landau orbits, and the values of the magnetic interaction parameters, λ_m, defined in Eq. (7.10). It thus follows that in order to sustain a critical field corresponding to $h = 10$ (H = 3050 gauss) at $t = 0.55$ ($T = 2.05$ K), we need to change v from unity (V = 3.51×10^{-19} ev cm^3) to about 1.02 (V = 3.59×10^{-19} ev cm^3), assuming that the Debye temperature remains the same. We also note that, for the case studied here, the magnetic interaction parameter has the value of 0.6 in the extreme quantum limit, which corresponds to $n_m(h\mathrm{H}_0) = 1$. This is similar to the value quoted in Ref. 16 in the same limit.

7.7.   de Haas–van Alphen Oscillations

Upon measuring the magnetization M of a sample of Bi as a function of the magnetic field H in high fields at 14.2 K in 1930, de Haas and van Alphen found oscillations in the M versus H plot. This effect is known as de Haas–van Alphen (dHvA) effect. Later, similar oscillatory behaviour was also found for the magnetic susceptibility of many elements as a function of H.

In the context of superconductivity, it was first pointed out in Ref. 10 — and corroborated by others e.g.,[11] — that $T_c(H)$ should exhibit dHvA effect in varying magnetic fields. We omit here the details (which are given in Ref. 18 below) of how our treatment of this effect differs from that of these authors, but note that the view advocated in these papers is that dHvA effect should be a characteristic of type-II SCs and not that of type-I. Not availing of this wisdom, and for the sake of concreteness, we choose to carry out here a study of Sn in the framework of BSE given above. This is done by proceeding as below.

We return to Eq. (7.21) and solve it for (a) t with h, v fixed and (b) h with t, v fixed. Value of v in each case is taken from Table 7.2. The results for Sn in case (a) for $h = 0.6$ ($v = 1.0269$) are shown in Fig. 7.3 in the plot of E (1.0269, 0.6, t) — see Eq. (7.21) — for an appropriate range of t. It crosses the x-axis only once, at $t = 0.632$, as expected. There are thus no oscillations in t for a fixed value of h. Figure 7.4 gives the plot for E (1.0269, h, 0.632) obtained by varying h (in steps of 0.001). In the range $0.58 \leq h \leq 0.62$, this curve crosses the zero-line about 20 times (if h is varied in steps of 0.0005, this number exceeds 40). From the corresponding values of $t(= \sqrt{1-h})$: $0.648 \geq t \geq 0.616$, and the H_c- and T_c-values of Sn (305 gauss, 3.72 K), it follows that a variation of H from 176.9 to 189.1 gauss yields about 20 solutions for T_c, ranging between 2.29 and 2.41 K. These are akin to dHvA oscillations of magnetic susceptibility in a varying field, and are caused by LQ which confines the electrons to a series of cylinders; in the absence of LQ, these electrons would be uniformly distributed throughout the Fermi sphere. As H, or h, is varied, the number of occupied cylinders (n_m) changes — vide Eq. (7.7), and this leads to a periodic dependence of various quantities on H. Such dependence for the parameters relevant to our problem is calculable because V $\neq$ 0 only in a specified range as given in Eq. (6.7); the results have been given in Table 7.2.

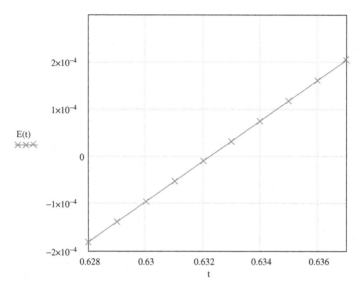

Fig. 7.3. Plot for Sn of the function E $(1.0269, 0.6, t)$ defined in Eq. (7.21), showing that for a fixed value of h, there is only one solution.

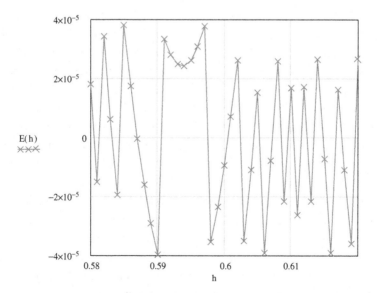

Fig. 7.4. dHvA oscillations exhibited by E $(1.0269, h, 0.632)$ — see Eq. (7.21) — as h is varied.

7.8.   Remarks

1. It is seen from Table 7.1 that ratio of the dimensionless parameters $\lambda_m(T = 0, H = H_c)$ and $\lambda(T = T_c, H = 0)$ is of the order of 10^{-3}. This result provides a *dynamical* explanation of the Meissner effect. This is so because both λ_m and λ were obtained by putting $|W| = 0$ (in their respective equations) which marks a transition from the super-conducting to the normal state. It therefore follows that placing an SC in an external magnetic field causes a significant reduction in the value of the interaction parameter responsible for pairing, and hence for the occurrence of superconductivity.

2. It was seen above that the equations for H_c of a non-elemental SC in the TPEM scenario involve *three* values of E_F — two for pairing caused via phonon-exchanges with each of the sub-lattices separately and one for pairing via simultaneous exchanges of phonons with both. In this connection, we draw attention to the "mountain" analogy discussed in Remark 5 in Chapter 4.

   Factually, E_Fs of HTSCs have of late been a subject of detailed inquiry via both theory[12,13] and experiment.[14–17] This is partly because, as was noted in Ref. 12, their low values are believed to be the cause of high-T_cs and partly because they shed light on the gap-structures of HTSCs. It seems interesting to note that on the basis of a detailed theoretical investigation, it has been reported in Ref. 13 that La_2CuO_4 should be characterized by the following three values of E_F: 60, 150, 240 meV. For the sake of comparison, we note that E_Fs of elemental SCs are in $\approx$2–10 eV range. The experimental work[14–17] concerned with the iron-pnictide SCs has revealed that their Fermi surfaces are characterized by several nodes or line-nodes.

3. Strictly speaking, the parameter V (as in [N(0)V]) should also have different values at different parts of the Fermi surface though, arguably, the variation in this parameter may not be as pronounced as for E_F. It was seen in Sec. 7.4 that V depends on T; so should E_F. This is a view that is fortified by a recent paper[18] where the T-dependence of the interaction parameter has been established via *experiment*.

4. It is seen from Eq. (7.7) that for a given value of θ_D, a change in H causes a change in $n_m(H)$, or the dHvA oscillations. This is akin to dependence of

the frequency of a monochord or sonometer on its length, tension, or mass per unit length. One is thus led to view dHvA oscillations as justifying the assumption of a cut off energy (Debye energy) for pairing in an SC. In other words, dHvA effect may not be thought of as a characteristic of type II SCs only.

Notes and References

1. J. Bardeen, L.N. Cooper and J.R. Schrieffer, *Phys. Rev.* **108**, 1175 (1957).
2. D. Shoenberg, *Superconductivity* (Cambridge University Press, 1965), p. 9.
3. C. Buzea and T. Yamashita, *Superconductors, Sci. Technol.* **14**, R115 (2001).
4. C.T. Rieck, K. Scharnberg and R.A. Klemm, *Physica C,* **170**, 195 (1990).
5. A.D. Linde, *Rep. Prog. Phys.* **42**, 389 (1979); A.A. Sokolov, I.M. Ternov, V. Ch. Zhukovskii, *Quantum Mechanics*, (Mir Publishers, Moskow, 1984), p. 261.
6. D. Pines, *Phys. Rev.* **109**, 280 (1958).
7. A.P. Cracknell and K.C. Kong, *The Fermi Surface* (Clarendon Press, Oxford, 1973).
8. L.W. Gruenberg and L. Gunther, *Phys. Rev.* **176**, 606 (1968).
9. Z. Tešanović, M. Rasolt and L. Xing, *Phys. Rev. Lett.* **63**, 2425 (1989); *Phys. Rev. B* **43**, 288 (1991).
10. L. Gunther and L.W. Gruenberg, *Solid State Commun.* **4**, 329 (1966).
11. A.K. Rajagopal and R. Vasudevan, *Phys. Lett.* **20**, 585 (1966); *Phys. Lett.* **23**, 539 (1966).
12. A.S. Alexandrov, http://arxiv.org/abs/cond-mat/0104413.
13. T. Jarlborg and A. Bianconi, *Phys. Rev. B* **87**, 054514 (2013).
14. H. Ding *et al.*, *Europhys. Lett.* **83**, 47001 (2008).
15. D. Lee, *Nature Phys.* **8**, 364 (2012).
16. Y. Zhang *et al.*, *Nature Phys.* **8**, 371 (2012).
17. M.P. Allan *et al.*, *Science* **336**, 563 (2012).
18. B. Mansart *et al.*, *Phys. Rev. B* **88**, 054507 (2013).
19. This chapter is based on: G.P. Malik, *Physica B* **405**, 3475 (2010).

Chapter 8

Dynamics-based Equations
for Critical Current Densities

8.1. Introduction

Critical current density (j_c) of an SC is the maximum current density that it can carry beyond which it loses the characteristic of superconductivity. It is an important parameter because greater its value, greater is the practical use to which the SC can be put. The *basic* relation between j_c and the critical velocity (v_c) of Cooper pairs (CPs) at any temperature T and an applied field H is:

$$j_c(T, H) = n_s(T, H)e^*v_c(T, H); \quad (v_c = P_c/2m^*), \quad (8.1)$$

where n_s is number of CPs, e^*, P_c and ($2m^*$) are, respectively, the charge, critical momentum, and the effective mass of a CP. We note that in BCS theory, CPs are visualized as existing interlocked with each other in the form of a diffused condensate. Therefore P_c in the above equation may also be defined as the minimum momentum that causes dissociation of the condensate into free electrons.

As alternatives to Eq. (8.1), several *derived* relations for j_c can be found in the literature.[1-4] Some of these relations have been reproduced in Table 8.1. Salient features of these expressions are: (a) they are based on diverse criteria such the type of SC being considered and its geometry, (b) they lead to values of j_cs that are generally higher than the experimental

113

Table 8.1.　Some of the relations obtained via diverse, indirect methods for calculating j_cs of different types of SCs.

S. No.	SC	j_c	Remark	Ref.
1	Type I; wire of radius a in the absence of external field	$H_c ac/2$; H_c: critical magnetic field c: velocity of light	j_c is the current that generates a field $= H_c$	
2	Type I; thin film or wire	$cH_c/4\pi\lambda$; λ: penetration depth	London theory; kinetic energy density is equated to condensation energy	1 p. 118
3	Type I; thin film or wire	$cH_c(T)/[3\sqrt{6}\pi\lambda(T)]$	Ginsburg–Landau theory	1 p. 117
4	Type I	$en_s\Delta/\hbar k_F$; e: electronic charge, n_s: density of superconducting electrons k_F: Fermi wave vector	BCS theory	2 p. 248
5	Type I; cylinder of radius a	$30\Delta\, M/a$; ΔM: width of magnetization loop at a given field and temperature	Bean's critical state model	3
6	Type I; slab of thickness d	$40\Delta\, M/d$	Bean's critical state model	3
7	Type I: hollow cylinder	$\alpha/(B + B_0)$; α and B_0 (thermodynamic critical magnetic field) are obtained from experiment	Kim et al. model	4

values, and (c) only one of them explicitly involves Fermi energy (via Fermi momentum) — this will be further discussed below.

In this chapter we present an approach to calculate $j_c(T)$ of an SC based directly on the dynamics of CPs. This is done by calculating $P_c(T)$, which is defined as the momentum at which the binding energy of the CPs vanishes (this is equivalent to the vanishing of the gap, as we know from Chapter 3). We are thus led to an expression for P_c/j_c that depends explicitly on E_F of the SC, besides depending directly on its familiar BCS parameters [N(0)V] and the Debye temperature θ_D. An advantage of the approach is that it is applicable regardless of the geometrical shape of the SC.

In the next section we obtain equations for the critical momentum of CPs in a simple SC. Note that this implies consideration of moving CPs, a feature that is easily incorporated into the theory via the BSE-based approach. Note also that in considering such CPs, we are going beyond the confines of the original BCS formalism which restricts the Hamiltonian at the outset to comprise of terms corresponding to pairs that have zero center-of-mass momentum. The equations for j_c obtained in Sec. 8.2 for elemental SCs are generalized for non-elemental SCs in Sec. 8.3.

8.2. Equations for $P_c(T)$ for an Elemental SC in the Absence of an External Magnetic Field

We derive in this section the T-dependent equation for the P_c of an elemental SC in the absence of any applied magnetic field. Our starting point is Eq. (2.3) in Chapter 2:

$$\left[\gamma_\mu^{(1)} P_\mu/2 + \gamma_\mu^{(1)} p_\mu - m + i\varepsilon\right]\left[\gamma_\mu^{(2)} P_\mu/2 - \gamma_\mu^{(2)} p_\mu - m + i\varepsilon\right]\chi(p_\mu)$$

$$= (-2\pi i)^{-1}\int d^4 q_\mu I(\mathbf{q} - \mathbf{p})\chi(q_\mu). \tag{8.2}$$

In all our work so far been, this equation has been dealt with by assuming that $P_\mu = (E, 0)$, i.e., for stationary CPs. Since we now wish to consider moving CPs, we have

$$P_\mu = (E, \mathbf{P}), \tag{8.3}$$

where $\mathbf{P}$ is the three-momentum of the center-of-mass of a CP in the lab frame. Multiplying Eq. (8.2) with $\gamma_4^{(1)}\gamma_4^{(2)}$ and simplifying as in Chapter 2, we now obtain

$$\gamma_4^{(1)}\left[\gamma_\mu^{(1)} P_\mu/2 + \gamma_\mu^{(1)} p_\mu - m + i\varepsilon\right]$$

$$= \gamma_4^{(1)}\left[\gamma_4^{(1)} E/2 - \boldsymbol{\gamma}^{(1)}\cdot\mathbf{P}/2 + \gamma_4^{(1)} p_4 - \boldsymbol{\gamma}^{(1)}\cdot\mathbf{p} - m + i\varepsilon\right]$$

$$= [E/2 + p_4 - H_1(\mathbf{p})],$$

where $H_1(\mathbf{p}) = (\mathbf{P}/2 + \mathbf{p}) \cdot \alpha^{(1)} + m\beta^{(1)}$. Dealing similarly with the expression in the second square brackets on the LHS of Eq. (8.2), we obtain

$$[E/2 + p_4 - H_1(\mathbf{p})][E/2 - p_4 - H_2(\mathbf{p})]\chi(p_\mu)$$

$$= (-2\pi i)^{-1} \int d^4 q_\mu I(\mathbf{q} - \mathbf{p})\chi(q_\mu), \quad (E = 2E_F + W) \quad (8.4)$$

which is identical with Eq. (2.4) in Chapter 2, *except* for the additional terms in the definitions of $H_1(\mathbf{p})$ and $H_2(\mathbf{p})$, where $H_2(\mathbf{p}) = (\mathbf{P}/2 - \mathbf{p}) \cdot \alpha^{(2)} + m\beta^{(2)}$. We now proceed as we did after obtaining Eq. (2.4) in Chapter 2. Subjecting Eq. (8.4) to the operation of $\Lambda_+^{(1)}\Lambda_+^{(2)}$, and defining

$$\left[\Lambda_+^{(1)}(\mathbf{p})\Lambda_+^{(2)}(\mathbf{p})\right]^{-1} [E/2 + p_4 - \Sigma_1(\mathbf{p})][E/2 - p_4$$

$$- \Sigma_2(\mathbf{p})]\chi_{++}(p_\mu) = S_{++}(\mathbf{p}),$$

we are led to

$$S_{++}(\mathbf{p}) = (-2\pi i)^{-1} \int d^3 q I(\mathbf{q} - \mathbf{p}) S_{++}(\mathbf{q}) \cdot J_{++}(\mathbf{q}), \quad (8.5)$$

where

$$J_{++}(\mathbf{q}) = \int \frac{dq_4}{[C(\mathbf{q}) + q_4][-q_4 + D(\mathbf{q})]}, \quad (8.6)$$

$$C(\mathbf{q}) = E/2 - (\mathbf{P}/2 + \mathbf{q})^2/2m = \left(E_F + \frac{W}{2} - \frac{\mathbf{P}^2}{8m} - \frac{\mathbf{P} \cdot \mathbf{q}}{2m} - \frac{\mathbf{q}^2}{2m}\right)$$

$$D(\mathbf{q}) = E/2 - (\mathbf{P}/2 - \mathbf{q})^2/2m = \left(E_F + \frac{W}{2} - \frac{\mathbf{P}^2}{8m} + \frac{\mathbf{P} \cdot \mathbf{q}}{2m} - \frac{\mathbf{q}^2}{2m}\right).$$

$$(8.7)$$

[see the discussion after Eq. (3.22) in Chapter 3].

We now apply Matsubara prescription to Eq. (8.6) by employing Eq. (2.12), whence

$$J_{++}(\mathbf{q}) = \frac{i\pi}{C(\mathbf{q}) + D(\mathbf{q})}[\tanh(\beta C(\mathbf{q})/2) + \tanh(\beta D(\mathbf{q})/2)]. \quad (8.8)$$

On substituting Eq. (8.8) into Eq. (8.5) and putting

$$\frac{[\tanh(\beta C(\mathbf{q})/2) + \tanh(\beta D(\mathbf{q})/2)]}{C(\mathbf{q}) + D(\mathbf{q})} S_{++}(\mathbf{q}) = \varphi(\mathbf{q}),$$

we have

$$\varphi(\mathbf{p}) = \left(\frac{-1}{2}\right) \frac{[\tanh(\beta C(\mathbf{p})/2) + \tanh(\beta D(\mathbf{p})/2)]}{C(\mathbf{p}) + D(\mathbf{p})} \int d^3 q I(\mathbf{q} - \mathbf{p})\varphi(\mathbf{q}). \tag{8.9}$$

In the spirit of BCS theory, we now choose for the kernel of this equation the following model

$$
\begin{aligned}
I(\mathbf{q} - \mathbf{p}) &= \frac{-V}{(2\pi)^3}(V > 0) \quad \text{for } E_F - k_B\theta_D \\
&\leq \frac{(\mathbf{P}/2 + \mathbf{p})^2}{2m}, \frac{(\mathbf{P}/2 - \mathbf{q})^2}{2m} \leq E_F + k_B\theta_D \\
&= 0, \quad \text{otherwise.}
\end{aligned} \tag{8.10}
$$

For this choice of the kernel, the pairing amplitude is a constant. On substituting Eq. (8.10) into Eq. (8.9) and multiplying with $\int d^3\mathbf{p}$, we obtain

$$1 = \frac{V}{16\pi^3} \int_L^U d^3\mathbf{p} \frac{[\tanh(\beta C(\mathbf{p})/2) + \tanh(\beta D(\mathbf{p})/2)]}{C(\mathbf{p}) + D(\mathbf{p})}, \tag{8.11}$$

where the limits L and U will be specified shortly.

Since $d^3 p = p^2 dp \sin\theta \, d\theta d\varphi$, in terms of $\xi = p^2/2m - E_F$, we have

$$p^2 dp = \frac{(2m)^{3/2} E_F^{1/2}}{2} d\xi, \text{ assuming that the range of integration} \ll E_F;$$

hence using the expressions for $C(\mathbf{p})$ and $D(\mathbf{p})$ given in Eq. (8.7), and noting that $\tanh(-x) = -\tanh(x)$, we can write Eq. (8.11) as:

$$1 = \frac{[N(0)V]}{8} \int_{-1}^1 dx \int_L^U d\xi \frac{\tanh[1] + \tanh[2]}{2(\xi - W/2 + \mathbf{P}^2/8m)}, \tag{8.12}$$

where $N(0)$ is recognized as the three-dimensional density of states (with proper factors of $\hbar$ and c inserted)

$$N(0) = \left[\frac{(2m)^{3/2} E_F^{1/2}}{4\pi^2}\right],$$

$$\tanh[1] = \tanh\left[\frac{\beta}{2}(\xi - W/2 + P\alpha x + P^2/8m)\right],$$

$$\tanh[2] = \tanh\left[\frac{\beta}{2}(\xi - W/2 - P\alpha x + P^2/8m)\right],$$

$$\alpha = \sqrt{E_F/2m}\ \left(p = \sqrt{2mE_F}\text{ has been used}\right), \quad \text{and}$$

$$x = \cos(\mathbf{P}, \mathbf{p}).$$

We now fix the limits L and U. The inequality in Eq. (8.10) can be written as

$$E_F - k_B\theta_D \leq \mathbf{p}^2/2m + P\alpha x + \mathbf{P}^2/8m,$$

$$\mathbf{p}^2/2m - P\alpha x + \mathbf{P}^2/8m \leq E_F + k_B\theta_D$$

$$\text{or,} \quad -k_B\theta_D - \mathbf{P}^2/8m \leq (\mathbf{p}^2/2m - E_F) + P\alpha x,$$

$$(\mathbf{p}^2/2m - E_F) - P\alpha x \leq k_B\theta_D - \mathbf{P}^2/8m$$

$$\text{or,} \quad -k_B\theta_D - \mathbf{P}^2/8m \leq \xi + P\alpha x,$$

$$\xi - P\alpha x \leq k_B\theta_D - \mathbf{P}^2/8m.$$

The requirement that $\xi + P\alpha x$ and $\xi - P\alpha x \geq -k_B\theta_D - \mathbf{P}^2/8m$ is automatically met if $\xi - P\alpha x \geq -k_B\theta_D - \mathbf{P}^2/8m$. Therefore L in (8.12) is fixed as

$$L = -k_B\theta_D + P\alpha x - \mathbf{P}^2/8m. \tag{8.13}$$

Similarly U is fixed as

$$U = k_B\theta_D - P\alpha x - \mathbf{P}^2/8m. \tag{8.14}$$

We now obtain the equation for $P_c(T)$ from Eq. (8.12) by putting $W = 0$ as

$$1 = \frac{[N(0)V]}{2}\int_0^1 dx \int_0^{k_B\theta_D - P_c\alpha x} d\xi$$

$$\times \frac{\tanh\left[\frac{\beta}{2}(\xi + P_c\alpha x)\right] + \tanh\left[\frac{\beta}{2}(\xi - P_c\alpha x)\right]}{\xi + P_c^2/8m}. \tag{8.15}$$

because the integrand is an even function of x; we have also assumed it to be an even function of ξ which is justified if $P_c\alpha x \ll k_B\theta_D$. This will turn out to be so. Since it also turns out that $P_c^2/8m \ll P_c \alpha x$, we have dropped it everywhere, excepting in the denominator in order to avoid the singularity at $\xi = 0$.

Equation (8.15) affords a consistency check of our procedure so far: If $P_c = 0$, the integral over x yields unity, the two tanh-functions add up and we obtain the correct BCS equation for T_c. In the next section we will solve Eq. (8.15) at $T = 0$ for Sn, for which results obtained by an alternative method are available in the literature.

8.3. Calculation of P_0 and the Number Density of Electrons for Sn From the Value of its j_0

Putting $[N(0)V] = \lambda$, we have Eq. (8.15) as

$$1 = \frac{\lambda}{2} \int_0^1 dx [I_1(x) + I_2(x)], \qquad (8.16)$$

where

$$I_1(x) = \int_0^{k_B \theta_D - P_c \alpha x} d\xi \frac{\tanh \left[\frac{\beta}{2}(\xi + P_c \alpha x)\right]}{\xi + P_c^2/8m},$$

$$I_2(x) = \int_0^{P_c \alpha x} d\xi \frac{\tanh \left[\frac{\beta}{2}(\xi - P_c \alpha x)\right]}{\xi + P_c^2/8m}$$

$$+ \int_{P_c \alpha x}^{k_B \theta_D - P_c \alpha x} d\xi \frac{\tanh \left[\frac{\beta}{2}(\xi - P_c \alpha x)\right]}{\xi + P_c^2/8m}.$$

The reason we have split Eq. (8.15) as above is: When $T = 0$ ($\beta = \infty$), $\tanh[] = 1$ in $I_1(x)$, whereas in $I_2(x)$ its value is (-1) for $\xi < P_c \alpha x$ and $(+1)$ for $\xi > P_c \alpha x$. It is convenient to similarly split $I_1(x)$. Denoting P_c by P_0 at $T = 0$, putting

$$E_1 = k_B \theta_D, \quad E_2 = P_0 \alpha, \quad \text{and} \quad E_3 = P_0^2/8m, \qquad (8.17)$$

we obtain

$$I_1(x) + I_2(x) = 2 \int_{P_0 \alpha x}^{k_B \theta_D - P_0 \alpha x} d\xi \frac{1}{\xi + P_0^2/8m}$$

$$= 2[\ln(E_1 + E_3 - E_2 x) - \ln(E_3 + E_2 x)].$$

On substituting this expression into Eq. (8.16) and carrying out the elementary integrations, we obtain

$$1 - \lambda \left[\frac{E_1 + E_3}{E_2} \ln \left(\frac{E_1 + E_3}{E_1 - E_2 + E_3} \right) + \ln \left(\frac{E_1 - E_2 + E_3}{E_1 + E_3} \right) \right.$$

$$\left. + \frac{E_3}{E_2} \ln \left(\frac{E_3}{E_2 + E_3} \right) \right] = 0. \tag{8.18}$$

Since, as will be seen below, $E_3 \ll E_1, E_2$, we can write this equation more compactly in terms of a dimensionless variable $y = E_1/E_2 = k_B\theta_D/P_0\alpha$ as

$$1 - \lambda \left[y \ln \left(\frac{y}{y-1} \right) + \ln(y-1) \right] = 0. \tag{8.19}$$

We now solve this equation for Sn for which the experimental values of T_c and θ_D are 3.72 K and 195 K, respectively. Therefore from the BCS relation $\lambda = -1/\ln(T_c/1.14\theta_D)$, we have $\lambda = 0.2445$.

Solving Eq. (8.19) with this value of λ, we obtain

$$y = \frac{k_B\theta_D}{P_0\alpha} = \frac{k_B\theta_D}{P_0} \sqrt{\frac{2m}{E_F}} = 22.48, \tag{8.20}$$

where the definition of α given after Eq. (8.12) has been used and $k_B\theta_D$, m, P_0, and E_F are in units of eV.

From Eqs. (8.1) and (8.20) we have

$$v_0 = \frac{P_0}{2m} = \frac{k_B\theta_Dc}{22.48\sqrt{2mc^2E_F}}, \tag{8.21}$$

where we have inserted the requisite factors of c so that it is manifestly seen that v_0 has the required dimensions. With $E_F = 1.74$ eV and the value of θ_D already given, Eq. (8.21) leads to

$$v_0 = 1.50 \times 10^4 \text{ cm sec}^{-1}. \tag{8.22}$$

In obtaining this value we have taken the effective mass of an electron[5] as $1.26 \times m_e$, where m_e is the free electron mass. Further, we have taken the value of E_F as 1.74 eV for the reason to be stated shortly. The value of critical current density j_0 for Sn is[2,6]

$$j_0 \approx 2 \times 10^7 \text{ Coulomb sec}^{-1}\text{ cm}^{-2}. \tag{8.23}$$

On substituting Eqs. (8.22) and (8.23) into (8.1), we obtain

$$n_s(\text{CPs}) = 4.17 \times 10^{21}\text{ cm}^{-3}; \quad n_s(\text{electrons}) = 8.34 \times 10^{21}\text{ cm}^{-3}. \tag{8.24}$$

For the first of these results, we have used for e* *twice* the value of the electronic charge. Before we proceed further, we note that using the above values of θ_D, m, E_F, and y in Eqs. (8.17)–(8.19) we obtain

$$E_1 = 1.68 \times 10^{-2}\text{ eV}, \quad E_2 = 7.48 \times 10^{-4}\text{ eV}, \quad E_3 = 8.03 \times 10^{-8}\text{ eV}$$

$$\frac{E_3}{E_2} \ln\left(\frac{E_3}{E_2 + E_3}\right) = -9.8 \times 10^{-4},$$

$$y \ln\left(\frac{y}{y-1}\right) = 1.023, \quad \ln(y-1) = 3.067. \tag{8.25}$$

These values justify the approximations made in obtaining Eq. (8.19).

We now compare the results above in Eqs. (8.22) and (8.24) with those that are obtained via an alternative approach as given in well-known text books.[2,6] In both these books, the equation invoked for j_c at $T = 0$ is:

$$j_0 = \frac{en_s\Delta_0}{p_F}, \tag{8.26}$$

where n_s is the number of electrons (not pairs) and p_F the Fermi momentum. This equation is equivalent to Eq. (8.1) because Δ/p_F has the dimension of velocity. The value of Fermi velocity used in these books is 6.97×10^7 cm/sec, which is equivalent to $E_F = 1.74$ eV. This is the reason we used this value above. With $\Delta_0 = 1.8 \times k_BT_c$, and $m = 1.26 \times m_e$ as earlier, Eq. (8.26) leads to

$$v_0 = \Delta/mv_F = 1.46 \times 10^4\text{ cm sec}^{-1}. \tag{8.27}$$

On using this value in Eq. (8.26) together with the value of j_0 given in Eq. (8.13), we obtain

$$n_s(\text{electrons}) = 8.50 \times 10^{21}\text{ cm}^{\times 3}. \tag{8.28}$$

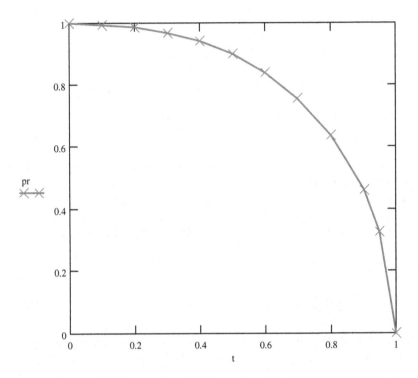

Fig. 8.1. Variation of the reduced critical momentum (p_r) of Sn with reduced temperature (t).

The results in Eqs. (8.27) and (8.28) are in excellent agreement with those in Eqs. (8.22) and (8.24), respectively.

Our considerations in this section so far have been limited to the $T = 0$ case. Since P_0 is known via Eq. (8.20), we can solve the T-dependent Eq. (8.15) in terms of the reduced (or normalized) variables $t = T/T_c$ and $p_r(t) = P_c(T)/P_0$. The plot of such solutions for Sn for $0 \leq t \leq 1$ is given in Fig. 8.1.

It is thus seen that the approach followed in this chapter yields results in accord with those obtained via an alternative, well-established indirect method. One may view our approach as providing *ab initio* justification for the latter approach. Additionally, it relates j_c with the relevant parameters of the problem at $T \neq 0$. In the next section we extend it to non-elemental SCs.

8.4. Equations for $P_c(T)$ and P_0 for a Non-elemental SC in the Absence of an External Magnetic Field

We consider here a non-elemental SC, $A_x B_{1-x}$, and seek to obtain generalized forms of Eqs. (8.15) and (8.18) when pairing is assumed to take place via TPEM. We now follow the procedure as was followed for such an SC in the same scenario in (a) Chapter 4 to obtain equations for the larger gap and T_c, and (b) Chapter 7 to obtain the equation for H_c. Specifically, we now choose for the kernel of Eq. (8.9) the following expression *in lieu* of Eq. (8.10):

$$
\begin{aligned}
I(\mathbf{q} - \mathbf{p}) = \frac{-(V_1^c + V_2^c)}{(2\pi)^3} \quad & (V_1^c > 0)V_1^c \neq 0 \quad \text{for } E_F - k_B \theta_D^{Ac} \\
& \leq (\mathbf{P}/2 + \mathbf{p})^2/2m, \quad (\mathbf{P}/2 - \mathbf{q})^2/2m \\
& \leq E_F + k_B \theta_D^{Ac} \\
& (V_2^c > 0)V_2^c \neq 0 \quad \text{for } E_F - k_B \theta_D^{Bc} \\
& \leq (\mathbf{P}/2 + \mathbf{p})^2/2m, \quad (\mathbf{P}/2 - \mathbf{q})^2/2m \\
& \leq E_F + k_B \theta_D^{Bc} \\
= 0, \quad \text{otherwise.} &
\end{aligned}
$$

(8.29)

In this expression, V_1^c is BCS model interaction parameter due to A ions in their combined state with B ions, and θ_D^{Ac} their Debye temperature in the same state, to be distinguished from V_1 and θ_D^A, respectively, which pertain to their free state.

On substituting Eqs. (8.29) into (8.9) and following the procedure that led to Eq. (8.15) — see also the brief discussion in Chapter 4 between Eqs. (8.9) and (8.10) — we obtain

$$
1 = \int_0^1 dx[J_1(x) + J_2(x)],
$$

(8.30)

where

$$J_1(x) = \frac{\lambda_1^c}{2} \int_0^{U_1} d\xi \frac{\psi(x, P_c, \xi)}{\xi + P_c^2/8m}, \quad J_2(x) = \frac{\lambda_2^c}{2} \int_0^{U_2} d\xi \frac{\psi(x, P_c, \xi)}{\xi + P_c^2/8m},$$

$$\psi(x, P_c, \xi) = \tanh\left[\frac{\beta}{2}(\xi + P_c\alpha x)\right] + \tanh\left[\frac{\beta}{2}(\xi - P_c\alpha x)\right]$$

$$U_1 = k_B\theta_D^{Ac} - P_c\alpha x, \quad U_2 = k_B\theta_D^{Bc} - P_c\alpha x$$

$$\tag{8.31}$$

On putting $T = 0$ in the above equations, we obtain the generalization of Eq. (8.18) to the TPEM scenario as:

$$1 - \lambda_1^c \Phi_1(\theta_D^{Ac}, \alpha, P_0) - \lambda_2^c \Phi_2(\theta_D^{Bc}, \alpha, P_0) = 0, \tag{8.32}$$

where

$$\Phi_1(\theta_D^{Ac}, \alpha, P_0) = \left[\frac{E_1 + E_3}{E_2} \ln\left(\frac{E_1 + E_3}{E_1 - E_2 + E_3}\right)\right.$$

$$+ \ln\left(\frac{E_1 - E_2 + E_3}{E_1 + E_3}\right) + \frac{P_0}{8m\alpha} \ln\left(\frac{P_0}{P_0 + 8m\alpha}\right)\bigg],$$

$$E_1 = k_B\theta_D^{Ac}, \quad E_2 = P_0\alpha, \quad E_3 = P_0^2/8m,$$

$$\tag{8.33}$$

and Φ_2 is obtained from Φ_1 by replacing E_1 by $k_B\theta_D^{Bc}$.

8.5.   Remarks

1. The equations obtained in this chapter can be generalized to include an external magnetic field by following the procedure given in the previous chapter.

2. To solve Eq. (8.33) for MgB$_2$, for example, we require as input $\lambda_{1,2}^c$, the two Debye temperatures and its E_F in the TPEM scenario. The last of these is not known. Note however that Eqs. (8.30) and (8.33) should be employed for MgB$_2$ only if the following inequality is satisfied.

$$E_F \gg k_B\theta_D^{Bc \text{ or } Mgc}. \tag{8.34}$$

3. Inequality Eq. (8.34) is a basic assumption of BCS theory. It is justified because E_Fs of elemental SCs are in $\approx$2–10 eV range whereas values of $k_B\theta_D$ for them are in the meV range. However, as was noted in Remark 1, Chapter 7, there are indications — both theoretical and experimental — that HTSCs are characterized by low E_F values. As a matter of fact there is a class of SCs for which E_Fs are so low as to be even *less* than their $k_B\theta_D$s. Among these are HFSCs. Because these SCs fall outside the purview of the conventional BCS theory, the term *unconventional* or *exotic* has been coined for them.

With the above property of HTSCs and HFSCs in view, it is natural to ask: Can one do away with constraint Eq. (8.34) and obtain for T_c and Δ equations that are valid for arbitrary values of E_F and $k_B\theta_D$? It is clear at the outset that in such equations E_F will be an additional independent variable. Therefore at least one more equation will be needed to supplement the equation for the T_c or Δ of the SC. We now call attention to the fact that this is precisely the situation that is met in the so-called BCS–BEC crossover physics. This is a topic that is usually not covered in the standard texts on superconductivity; we deal with it in some detail in the next chapter because it sets the stage for dealing with real-life SCs like $SrTiO_3$ and HFSCs in a manner not attempted before.

Notes and References

1. M. Tinkham, *Introduction to Superconductivity* (McGraw Hill, NY, 1975).
2. H. Ibach and H. Lüth, *Solid State Physics* (Springer, Berlin, 1996).
3. C.P. Bean, *Phys. Rev. Lett.* **36**, 31 (1964).
4. Y.B. Kim, C.F. Hempstead, and A.R. Strand, *Phys. Rev.* **129**, 528 (1963).
5. C. Kittel, *Introduction to Solid State Physics* (Wiley Eastern, New Delhi, 1974).
6. A.C. Rose-Innes and E.H. Rhoderic, *Introduction to Superconductivity* (Pergamon, Oxford, 1978)
7. This chapter is based on: G.P. Malik, *WJCMP* **3**, 103 (2013).

Chapter 9

BCS–BEC Crossover Physics Without Appeal to Scattering Length Theory

9.1. Introduction

9.1.1. *The need to incorporate E_F into the equations for $\Delta(T)$ and T_c*

We recall that E_F does not figure in the equation for Δ, or T_c, as it is usually written because it is assumed in BCS theory that

$$E_F(\text{or } \mu) \gg \hbar\omega_c = k_B\theta_D, \tag{9.1}$$

where the symbols have their usual significance. It was noted in Remark 3 in the last chapter that there are indications that constraint Eq. (9.1) is not satisfied by many SCs. Besides, T_cs of some SCs have been known for a long time to depend in a marked manner on their electron concentration n, SrTiO$_3$ being a prime example. Therefore, since n is related with E_F, there would seem to be a need to obtain equations for Δ and T_c that include E_F as a variable and are not constrained by Eq. (9.1). Such equations are in fact an integral part of BCS theory. Their application to dilute SrTiO$_3$ by Eagles[1] perhaps marks the beginning of what is now known as BCS–BEC crossover physics.

9.1.2.   The meaning of crossover

It is generally believed that $T = T_c$ marks the temperature at which formation of Cooper pairs (CPs) begins and superconductivity arises.[2] At this
temperature, $\Delta = 0$; thereafter, as T is decreased, $\Delta(T)$ progressively
increases till it attains its maximum value at $T = 0$. As was shown in
Chapter 3, $2|\Delta(T)| = 2|W(T)|$ around the Fermi surface signifies the
region of pair-formation. Since the total energy of a CP at $T = 0$ is
$2E_F + |W_0| \approx 2F_F$ because $E_F \gg |W_0|$, the momentum corresponding to it is $\Delta p = 2E_F/c \approx 2.7 \times 10^{-10}$ cgs units, for the average value of
E_F taken as 4 eV. This implies via the uncertainty principle that the spatial extent of a CP at $T = 0$ is of the order of 1.5×10^{-5} cm, i.e., about
three orders of magnitude greater than the size of a hydrogen atom. Since
this is also the region of pair-formation, it follows that within the spatial
extent of a CP lie the centres of many millions and more of other CPs.
Hence CPs may not be thought of as independent particles, but as a *diffused
aggregate* to which the interlocking of its constituents lends stability. It is
in this sense that BCS state at any temperature is spoken of as a *condensate*
of CPs.

The word *condensate* is evocative of Bose–Einstein condensation (BEC)
— a phenomenon for which theory preceded experiment by about 70 years,
unlike superconductivity for which experiment preceded theory by about 46
years. There are of course differences between the two phenomena: While
BEC is about condensation of bosons, superconductivity is about condensation of CPs, i.e., composites of electrons which are fermions. In fact
it was categorically stated in BCS[3] (Footnote 18) that "*our transition* (to
superconducting state) *is not analogous to a Bose–Einstein condensation.*"
Nonetheless superconductivity and BEC have been viewed as two sides
of the same coin since the time of the quasi-chemical equilibrium theory
of superconductivity given by Schafroth and collaborators.[4] Eagles'[1] study,
concerned with thin films of $SrTiO_3$ mentioned earlier and carried out in the
framework of BCS theory, was a definitive step contributing to such a view.
Later, Leggett[5] revisited the equations considered by Eagles by employing the framework of the scattering length theory (SLT) in the low energy
domain.

9.1.3. *Salient features of crossover via SLT*

By recasting the equation that governs collisions between electrons and leads to the BCS gap equation in terms of the scattering length a, and invoking the BCS number equation, Leggett showed that: (i) when $\xi \equiv 1/k_F a \to -\infty$ ($k_F \equiv$ Fermi wave vector) and $\mu/E_F \to 1$ ($\mu \equiv$ chemical potential, $E_F \equiv$ Fermi energy), the gap Δ is exponentially small; this is the extreme BCS regime in which a CP has a rather large size within the domain of which lie the centers-of-mass of very many other pairs and (ii) the limits $\xi \to +\infty$ and $\mu \to -\hbar^2/2ma^2$ ($m \equiv$ mass of an electron) lead to shrinkage in the size of a pair till it resembles a tightly bound molecule far apart from other similarly bound pairs.

While Leggett's[5] work at temperature $T = 0$ dealt with the two extremes values of ξ noted above, it claimed to be valid for arbitrary values of ξ — implying thereby continuity of the plots of Δ and μ against ξ between the two limits. This is a feature that has been validated via both analytic[6,7] and numerical solutions[8] of the two coupled equations involving Δ and μ. Leggett's analysis of the problem at $T = 0$ was extended to models other than the BCS model and to non-zero temperatures by Nozières and Schmitt-Rink[9] who pointing out that while the transition temperature T_c in the BCS regime is controlled by ionization (i.e., breaking up) of the pairs, it is controlled by the centre-of-mass motion of the pairs in the BEC regime and that it too varies continuously from a weak BCS regime to a strong BEC regime. The literature on the crossover problem is huge; for further references we refer the reader to the review by Chen *et al.*[10] who discuss the intersection of high-T_c superconductivity and superfluidity in ultracold fermionic atomic gases from the crossover viewpoint.

9.1.4. *Scope of this chapter*

Post mid-1980s, following Leggett, most of the work on the crossover problem seems to have been done in the SLT framework. Since the limits of the integrals in the equations for Δ and μ in this approach are from zero to infinity, one needs the not-so-familiar "matrix-algebraic tricks"[5,11] to regularize or renormalize the equations. While $\xi \to \infty$ limit in Leggett's work, and in the papers following it up, is generally regarded as the signifying the extreme of the BEC regime, it was stressed in Ref. 5 that CP must be regarded as *"automatically Bose-condensed,"* because their (molecular) wave function is *"common to all N/2 pairs."* Be that as it may, what seems missing in these papers is a definitive, *quantitative* signature of BEC taking place, such as a relation akin to $\rho \, (\lambda_{dB})^3 \geq 2.612$ ($\rho =$ density; $\lambda_{dB} =$ de Broglie wavelength) being appealed to and shown to be valid.

We present in this chapter a study of crossover physics without appeal to SLT. By progressively decreasing the number density n (or E_F), we first solve the usual BCS equations for Δ_0 and the chemical potential μ_0 that Eagles had considered[1]; this exercise is then repeated by replacing the equation for Δ_0 by the equivalent equation for $|W_0|$ obtained via BSE-based approach in Chapter 3. In either case, the cut offs that BCS theory provides suffice to obtain finite results without the "tricks" that the SLT approach requires. We are thus enabled to give, as it were, a frame by frame picture of the crossover phenomenon. Specifically we show, as expected, that plots of Δ_0 and μ_0 obtained via the usual BCS equations are very similar to the plots obtained by solving the equation for $|W_0|$ and the number equation in which Δ_0 is replaced by $|W_0|$.

The advantages of the BSE-based approach are: (a) it sheds light on a feature of crossover that has hitherto not been addressed: The role of the hole–hole scatterings *vis-à-vis* the electron-electron scatterings as one goes from the BCS to the BEC end and, more importantly, (b) it leads to critical values of the concentration (n) and the interaction parameter (λ) at which the *diffused* condensate of CPs acquires a *tangible* character akin to those of BE condensates[12] that have been observed since the mid-1990s. Our approach also leads to an estimate of the number of CPs which congregates to form a well-defined macro state of the BE condensate. This is in accord with the contention that BEC of electrons is a 2-step process — formation of pairs into a *diffused* aggregate and *then* their conversion into a *well-defined* BE-condensed state. Finally, our treatment — which differs from that of Eagles[1] — will be shown in the next chapter to shed new light on the puzzle posed by $SrTiO_3$. In a subsequent chapter, it will also be shown to shed new light on the superconducting features of La_2CuO_4 and the so-called unconventional or exotic HFSCs.

9.2. Framework of Crossover: The Usual BCS Equations

Since we need to progressively decrease E_F for our study, we now need to do away with constraint Eq. (9.1). To this end we begin with the gap and the number-density equations as

$$1 = \frac{V}{2} \int_{\mu-\omega_c}^{\mu+\omega_c} \frac{d^3p}{(2\pi)^3} \frac{1}{[(p^2/2m - \mu)^2 + \Delta^2]^{1/2}},$$

$$n = \int_0^{\mu+\omega_c} \frac{d^3p}{(2\pi)^3} \left[1 - \frac{(p^2/2m - \mu)}{[(p^2/2m - \mu)^2 + \Delta^2]^{1/2}} \right], \qquad (9.2)$$

where V is the measure of the net attraction between a pair of electrons. In terms of $\xi = (p^2/2m - \mu)$, we have Eqs. (9.2) as

$$1 = \frac{1}{2} V \left[\frac{(2m)^{3/2}}{4\pi^2} \right] \int_{-\omega_c}^{\omega_c} \frac{d\xi(\xi+\mu)^{1/2}}{[\xi^2 + \Delta^2]^{1/2}},$$

$$n = \left[\frac{(2m)^{3/2}}{4\pi^2} \right] \int_{-\mu}^{\omega_c} d\xi(\xi+\mu)^{1/2} \left[1 - \frac{\xi}{(\xi^2 + \Delta^2)^{1/2}} \right]. \qquad (9.3)$$

Multiplying the first of these equations by $E_F^{1/2}$ and using the relations

$$\left[\frac{(2m)^{3/2}}{4\pi^2} \right] E_F^{1/2} = N(E_F), \quad \lambda = N(E_F)V, \quad n = \left[\frac{(2mE_F)^{3/2}}{3\pi^2} \right],$$

$$(\text{N}(E_F) \equiv \text{the density of states at the Fermi surface}) \qquad (9.4)$$

the set of Eqs. (9.3) may be written as

$$E_F^{1/2} = \frac{1}{2} \lambda \int_{-\omega_c}^{\omega_c} \frac{d\xi(\xi+\mu)^{1/2}}{[\xi^2 + \Delta^2]^{1/2}},$$

$$\frac{4}{3} E_F^{3/2} = \int_{-\mu}^{\omega_c} d\xi(\xi+\mu)^{1/2} \left[1 - \frac{\xi}{[\xi^2 + \Delta^2]^{1/2}} \right]. \qquad (9.5)$$

We now have the desired equations with E_F-dependence made explicit. It is convenient however to choose E_F as the *unit of energy* and write these equations in terms of scaled variables defined as

$$\tilde{\xi} = \xi/E_F, \quad \tilde{\omega}_c = \omega_c/E_F, \quad \tilde{\mu} = \mu/E_F, \quad \tilde{\Delta} = \Delta/E_F, \qquad (9.6)$$

whence we have

$$1 = \frac{1}{2}\lambda \int_{-\tilde{\omega}_c}^{\tilde{\omega}_c} \frac{(\tilde{\xi} + \tilde{\mu})^{1/2} d\tilde{\xi}}{[\tilde{\xi}^2 + \tilde{\Delta}^2]^{1/2}},$$

$$\frac{4}{3} = \int_{-\tilde{\mu}}^{\tilde{\omega}_c} d\tilde{\xi} \, (\tilde{\xi} + \tilde{\mu})^{1/2} \left[1 - \frac{\tilde{\xi}}{[\tilde{\xi}^2 + \tilde{\Delta}^2]^{1/2}} \right]. \qquad (9.7)$$

Defining

$$f_1(\tilde{\Delta}, \tilde{\mu}, \tilde{\xi}) = \frac{(\tilde{\xi} + \tilde{\mu})^{1/2}}{[\tilde{\xi}^2 + \tilde{\Delta}^2]^{1/2}},$$

$$f_2(\tilde{\Delta}, \tilde{\mu}, \tilde{\xi}) = (\tilde{\xi} + \tilde{\mu})^{1/2} \left[1 - \frac{\tilde{\xi}}{[\tilde{\xi}^2 + \tilde{\Delta}^2]^{1/2}} \right]$$

we now write Eqs. (9.7) as

$$1 = \frac{\lambda}{2}\left[I_{11}(\tilde{\Delta}, \tilde{\mu}) + I_{12}(\tilde{\Delta}, \tilde{\mu}) \right], \quad \frac{4}{3} = I_{21}(\tilde{\Delta}, \tilde{\mu}) + I_{22}(\tilde{\Delta}, \tilde{\mu}), \quad (9.8)$$

where

$$I_{11}(\tilde{\Delta}, \tilde{\mu}) = \int_{-\tilde{\omega}_c}^{0} f_1(\tilde{\Delta}, \tilde{\mu}, \tilde{\xi}) d\tilde{\xi},$$

$$I_{12}(\tilde{\Delta}, \tilde{\mu}) = \int_{0}^{\tilde{\omega}_c} f_1(\tilde{\Delta}, \tilde{\mu}, \tilde{\xi}) d\tilde{\xi} \qquad (9.9)$$

$$I_{21}(\tilde{\Delta}, \tilde{\mu}) = \int_{-\tilde{\mu}}^{0} f_2(\tilde{\Delta}, \tilde{\mu}, \tilde{\xi}) d\tilde{\xi},$$

$$I_{22}(\tilde{\Delta}, \tilde{\mu}) = \int_{0}^{\tilde{\omega}_c} f_2(\tilde{\Delta}, \tilde{\mu}, \tilde{\xi}) d\tilde{\xi}. \qquad (9.10)$$

Equations (9.9) enable us to introduce

$$\rho(\tilde{\Delta}, \tilde{\mu}) = I_{11}(\tilde{\Delta}, \tilde{\mu})/I_{12}(\tilde{\Delta}, \tilde{\mu}), \qquad (9.11)$$

which is seen to be the ratio of the contribution of the hole–hole scatterings and the electron-electron scatterings to the pairing amplitude as n is varied.

Finally, we also define

$$R \equiv (2|\mu| + \Delta)/2E_F = |\tilde{\mu}| + \tilde{\Delta}/2, \qquad (9.12)$$

which, as will be shown below, sheds light on the values of n and λ that mark the onset of BEC for the BCS model interaction.

The framework for our study of the crossover problem via the framework of the usual BCS equations is now complete. In the next section we visit the crossover problem via Eqs. (9.8). In the section following it, we revisit the problem via the alternative equations discussed in Sec. 1. We are thus enabled to (i) obtain a frame-by-frame picture, as it were, of the transition to the BEC regime starting from a well-understood, manifestly BCS state (ii) unravel a feature of the crossover — via the parameter ρ defined in Eq. (9.11) — that has not been under the purview of the approach based on SLT, and (iii) identify the values of λ and n (or E_F) that mark the onset of a macroscopic occupation of a paired state characteristic of BEC.

9.3. Crossover via Eqs. (9.8)

For the sake of concreteness let us consider the hypothetical case of Sn, the superconducting parameters of which are:

$$\omega_c = k_B\theta_D = 0.0168\,\text{eV}(\theta_D = 195\text{ K}), \quad \Delta_0 = 5.93 \times 10^{-4}\,\text{eV},$$
$$T_c = 3.72\text{ K}, \quad \lambda = 0.2477; \qquad (9.13)$$

we also have

$$E_F(\text{Sn}) = \frac{1}{2m}(3\pi^2 n)^{2/3} = 10.055\,\text{eV}, \qquad (9.14)$$

since $n = $ atomic concentration (3.62×10^{22}) × no. of valence electrons (4) $= 1.448 \times 10^{23}$.

Solving Eqs. (9.8) with the input of ω_c, λ and E_F from the above equations, we obtain $\tilde{\Delta} = 5.9 \times 10^{-5}$ and $\tilde{\mu} = 1$; multiplying these with the scaling factor, i.e., 10.055 eV, we obtain the value for Δ_0 as in Eq. (9.13) and $\mu = E_F$ as in Eq. (9.14). Thus, for $\lambda = 0.2477$, we have a well-understood, manifestly BCS state. In the following, for each value of n (or E_F) that we consider, we let λ vary between 0.1 and 0.9 (this will be further discussed below) in order to unravel the salient features of the crossover.

We begin with $n = 1.448 \times 10^{23}$ cm^{-3}, which determines E_F via Eq. (9.4) and hence $\tilde{\omega}_c$ via Eqs. (9.13) and (9.6). For each of the values of λ noted above, we now solve Eqs. (9.8) for $\tilde{\Delta}$ and $\tilde{\mu}$. This exercise is repeated by reducing n in steps by a factor of 10 and going down up to 1.448×10^{14} cm^{-3}. We note that n can be decreased via both a dilution of the atomic concentration and a reduction in the number of valence electrons from 4 to 3, 2, or 1. While it is expected that V (and hence λ) should increase as the concentration is lowered because it leads to a weakening of the repulsive interaction between electrons, the precise quantum of change in V remains unknown. We recall that a rather detailed model for the variation of λ with n was given by Eagles[13] in the context of SrTiO$_3$, which brings out the complexity of the situation as the experimental features are varied. It thus seems reasonable to assume that changing n may cause λ to have any value <1 for the reason to be discussed below. The results of our calculations are presented in Table 9.1 While variation of the single parameter n is akin to variation of the scattering length in the customary approach, the additional features of the present approach are that it also sheds light on the parameters ρ and R defined above.

It is worth remarking that the crossover picture that we are led to is not dependent on our choice of Sn because a reduction in n of any other superconducting element/compound will cause changes in the values of its E_F and $\tilde{\omega}_c$ similar to those that we have described for Sn. We chose to make Sn a vehicle for our study because it has *four* valence electrons (one or more of which can be "sucked away" — by whatever means); one thus has an additional channel, besides dilution of the atomic concentration, for a reduction in the number of conduction electrons.

9.4. Procedural Details About Calculations

It is seen from Table 9.1 that $\tilde{\omega}_c$ — the limiting value of the integrals in Eqs. (9.9) and (9.10) — increases as n or E_F is decreased. So long as $\tilde{\omega}_c$ remains less than $\tilde{\mu}$, i.e., up to $E_F = 2.17 \times 10^{-1}$ eV ($\tilde{\omega}_c = 0.776$), one can solve the set of Eq (9.8) without any problem.

At $E_F = 4.67 \times 10^{-3}$ eV ($\tilde{\omega}_c = 3.6$), the limits of the integral $I_{11}(\tilde{\Delta}, \tilde{\mu})$ in Eqs. (9.9) are from -3.6 to 0 whereas $\tilde{\mu}$ is expected to be of order unity or less, whence the factor $(\tilde{\xi} + \tilde{\mu})^{1/2}$ in the integrand can become imaginary

Table 9.1. Values of $\tilde{\Delta}$ and $\tilde{\mu}$ obtained via the set of coupled equations (8) as λ is varied from 0.1 to 0.9 for different values of n/E_F. The values of ρ corresponding to these solutions are obtained via Eq. (9.11).

n (cm^{-3})	1.448×10^{14}		1.448×10^{15}		1.448×10^{16}		1.448×10^{17}		1.448×10^{18}		1.448×10^{19}		1.448×10^{20}		1.448×10^{21}		1.448×10^{22}		1.448×10^{23}	
E_F (eV)	1.01×10^{-5}		4.67×10^{-5}		2.17×10^{-4}		1.01×10^{-3}		4.67×10^{-3}		2.17×10^{-2}		1.01×10^{-1}		4.67×10^{-1}		2.17		10.1	
$\bar{\omega}_c$	1.67×10^{3}		360.04		77.57		16.71		3.6		7.76×10^{-1}		1.67×10^{-1}		3.6×10^{-2}		7.76×10^{-3}		1.67×10^{-3}	
λ	$\tilde{\Delta}$	$\tilde{\mu}, \rho$	$\tilde{\Delta}$	$\tilde{\mu}, \rho$	$\tilde{\Delta}$	$\tilde{\mu}, \rho$	$\tilde{\Delta}$	$\tilde{\mu}, \rho$	$\tilde{\Delta}$	$\tilde{\mu}, \rho$	$\tilde{\Delta}$	$\tilde{\mu}, \rho$	$\tilde{\Delta}$	$\tilde{\mu}, \rho$	$\tilde{\Delta}$	$\tilde{\mu}, \rho$	$\tilde{\Delta}$	$\tilde{\mu}, \rho$	$\tilde{\Delta}$	$\tilde{\mu}, \rho$
0.1	12.06	$-599.4, 0$	4.34	$-41.01, 0$	2.79×10^{-1}	$0.93, 0.07$	2.59×10^{-3}	$1, 0.43$	2.53×10^{-4}	$1, 0.72$	6.76×10^{-5}	$1, 0.92$	1.52×10^{-5}	$1, 0.98$	3.29×10^{-6}	$1, 1$	7.08×10^{-7}	$1, 1$	1.53×10^{-7}	$1, 1$
0.15	15.72	$-802.2, 0$	6.29	$-86.92, 0$	1.77	$-1.87, 0$	7.18×10^{-2}	$0.99, 0.26$	7.10×10^{-3}	$1, 0.61$	1.89×10^{-3}	$1, 0.89$	4.25×10^{-4}	$1, 0.98$	9.16×10^{-5}	$1, 0.99$	1.97×10^{-5}	$1, 1$	4.25×10^{-6}	$1, 1$
0.2	18.72	$-929.9, 0$	7.80	$-120.6, 0$	2.72	$-7.07, 0$	3.40×10^{-1}	$0.90, 0.14$	3.75×10^{-2}	$1, 0.51$	1.00×10^{-2}	$1, 0.85$	2.25×10^{-3}	$1, 0.97$	4.85×10^{-4}	$1, 0.99$	1.05×10^{-4}	$1, 1$	2.25×10^{-5}	$1, 1$
0.3	23.68	$-1084.9, 0$	10.22	$-165.1, 0$	4.04	$-16.71, 0$	1.12	$0.18, 0.01$	1.91×10^{-1}	$0.97, 0.36$	5.30×10^{-2}	$1, 0.80$	1.19×10^{-2}	$1, 0.95$	2.57×10^{-3}	$1, 0.99$	5.54×10^{-4}	$1, 1$	1.19×10^{-4}	$1, 1$
0.4	27.81	$-1178, 0$	12.20	$-193.4, 0$	5.04	$-23.98, 0$	1.73	$-0.90, 0$	4.12×10^{-1}	$0.89, 0.26$	1.22×10^{-1}	$0.99, 0.76$	2.76×10^{-2}	$1, 0.94$	5.95×10^{-3}	$1, 0.99$	1.28×10^{-3}	$1, 1$	2.76×10^{-4}	$1, 1$
0.5	31.42	$-1240, 0$	13.92	$-213.2, 0$	5.90	$-29.47, 0$	2.21	$-1.97, 0$	6.39×10^{-1}	$0.78, 0.19$	2.02×10^{-1}	$0.98, 0.73$	4.61×10^{-2}	$1, 0.94$	9.93×10^{-3}	$1, 0.99$	2.14×10^{-3}	$1, 1$	4.61×10^{-4}	$1, 1$
0.6	34.68	$-1286, 0$	15.47	$-228, 0$	6.65	$-33.74, 0$	2.61	$-2.92, 0$	8.48×10^{-1}	$0.66, 0.13$	2.85×10^{-1}	$0.98, 0.71$	6.55×10^{-2}	$1, 0.94$	1.41×10^{-2}	$1, 0.99$	3.04×10^{-3}	$1, 1$	6.55×10^{-4}	$1, 1$
0.7	37.67	$-1321, 0$	16.88	$-239.5, 0$	7.33	$-37.17, 0$	2.96	$-3.75, 0$	1.04	$0.53, 0.09$	3.69×10^{-1}	$0.97, 0.69$	8.52×10^{-2}	$1, 0.93$	1.83×10^{-2}	$1, 0.98$	3.95×10^{-3}	$1, 1$	8.50×10^{-4}	$1, 1$
0.8	40.44	$-1350, 0$	18.19	$-248.7, 0$	7.96	$-39.99, 0$	3.28	$-4.47, 0$	1.21	$0.41, 0.06$	4.51×10^{-1}	$0.96, 0.68$	1.05×10^{-1}	$1, 0.93$	2.25×10^{-2}	$1, 0.98$	4.84×10^{-3}	$1, 1$	1.04×10^{-3}	$1, 1$
0.9	43.04	$-1373, 0$	19.41	$-256.3, 0$	8.55	$-42.36, 0$	3.58	$-5.10, 0$	1.37	$0.30, 0.04$	5.31×10^{-1}	$0.95, 0.67$	1.24×10^{-1}	$1, 0.93$	2.67×10^{-2}	$1, 0.98$	5.73×10^{-3}	$1, 1$	1.23×10^{-3}	$1, 1$

over a part of the integration interval. Therefore, in order to obtain real solutions of our problem, we must change the lower limit in the first of Eq. (9.9) from $-\tilde{\omega}_c$ to $-\tilde{\mu}$. For this value of E_F, no other change is required.

At $E_F = 1.01 \times 10^{-3} (\tilde{\omega}_c = 16.71)$, we can continue as above up to $\lambda = 0.3$ for which the solutions are: $\tilde{\Delta} = 1.12$, $\tilde{\mu} = 0.18$. In going over to $\lambda = 0.4$ now, we find that for the first time $\tilde{\mu}$ becomes **negative**, which causes $I_{11}(\tilde{\Delta}, \tilde{\mu})$ and $I_{21}(\tilde{\Delta}, \tilde{\mu})$ to become pure imaginary, and $I_{12}(\tilde{\Delta}, \tilde{\mu})$ and $I_{22}(\tilde{\Delta}, \tilde{\mu})$ to become complex. Whenever this happens, real solutions are obtained by restricting the limits of the gap and the number equations to the interval $\{-\tilde{\mu}, \tilde{\omega}_c\}$. Effectively, this means that we must now set $I_{11}(\tilde{\Delta}, \tilde{\mu}) = I_{21}(\tilde{\Delta}, \tilde{\mu}) = 0$ and take the lower limits for $I_{21}(\tilde{\Delta}, \tilde{\mu})$ and $I_{22}(\tilde{\Delta}, \tilde{\mu})$ as $-\tilde{\mu}$.

9.5.　Crossover via Alternative Equations

It is seen from Table 9.1 that for values of $E_F \geq 4.67 \times 10^{-3}$ eV the values of $\tilde{\Delta}$ are greater than unity for certain values of λ. For an interpretation of these, we revisit the crossover problem via an alternative to the gap equation dealt with above. The alternative equation is obtained from Eq. (3.24), which was discussed in detail in Chapter 3:

$$1 = \frac{V}{(2\pi)^3} \frac{1}{2} \int_{E_F - k_B \theta_D}^{E_F + k_B \theta_D} d^3\mathbf{p} \frac{\tanh[(\beta/2)(\mathbf{p}^2/2m - E_F - W/2)]}{\mathbf{p}^2/2m - E_F - W/2}, \quad (9.15)$$

where $\beta = 1/k_B T$ (k_B being the Boltzmann constant), and W is defined via

$$\text{E (total energy of a CP)} = 2E_F + W. \quad (9.16)$$

Since we intend to beyond the $E_F = \mu$ scenario, we write Eq. (9.15) as

$$1 = \frac{V}{4\pi^2}[I_1 + I_2], \quad (9.17)$$

where, with

$$F(p^2) = \frac{p^2 \tanh[\frac{\beta}{2}(p^2/2m - \mu - W/2)]}{(p^2/2m - \mu - W/2)}, \quad (9.18)$$

$$I_1 = \int_{\mu - \omega_c}^{\mu} F(p^2)dp, \quad I_2 = \int_{\mu}^{\mu + \omega_c} F(p^2)dp. \quad (9.19)$$

We first deal with I_2, which pertains to the region of electron–electron scatterings. As explained in Chapter 3, we assume that in this region $W =$

$-|W|$. Then, with the definition $\xi = p^2/2m - \mu$, we have

$$I_2 = \frac{(2m)^{3/2}}{2} \int_0^{\omega_c} \frac{(\xi + \mu)^{1/2} \tanh[\frac{\beta}{2}(\xi + |W|/2)]d\xi}{(\xi + |W|/2)},$$

which, in the limit $\beta \to \infty$ (T $\to$ 0), reduces to

$$I_2 = \frac{(2m)^{3/2}}{2} \int_0^{\omega_c} \frac{(\xi + \mu)^{1/2}d\xi}{(\xi + |W|/2)}. \tag{9.20}$$

We now deal with I_1, which corresponds to the region of hole–hole scatterings, by assuming that $W = |W|$, whence

$$I_1 = \frac{(2m)^{3/2}}{2} \int_{-\omega_c}^0 \frac{(\xi + \mu)^{1/2} \tanh\left[\frac{\beta}{2}(\xi - |W|/2)\right] d\xi}{(\xi - |W|/2)}.$$

The tanh now tends to (-1) in the $\beta \to \infty$ limit since $\xi \leq 0$, whence

$$I_1 = -\frac{(2m)^{3/2}}{2} \int_{-\omega_c}^0 \frac{(\xi + \mu)^{1/2}d\xi}{(\xi - |W|/2)}. \tag{9.21}$$

Substituting Eqs. (9.20) and (9.21) into Eq. (9.17), multiplying the resulting equation by $E_F^{1/2}$, using Eq. (9.4) and then carrying out scaling as in Eq. (9.6), we obtain

$$1 = \frac{\lambda}{2}[J_{11}(|\tilde{W}|, \tilde{\mu}) + J_{12}(|\tilde{W}|, \tilde{\mu})], \tag{9.22}$$

where $|\tilde{W}| = |W|/E_F$ and

$$J_{11}(|\tilde{W}|, \tilde{\mu}) = \int_{-\omega_c}^0 \frac{(\tilde{\xi} + \tilde{\mu})^{1/2}d\tilde{\xi}}{-\tilde{\xi} + |\tilde{W}|/2},$$

$$J_{12}(|\tilde{W}|, \tilde{\mu}) = \int_0^{\omega_c} \frac{(\tilde{\xi} + \tilde{\mu})^{1/2}d\tilde{\xi}}{\tilde{\xi} + |\tilde{W}|/2}. \tag{9.23}$$

We now solve Eq. (9.22) and the second of Eqs. (9.7) (with $\tilde{\Delta}$ replaced by $|\tilde{W}|$). Procedural details for obtaining these solutions are similar to those given in Sec. 9.4. Remarkably, we find that for all values of n and λ, the solutions for $|\tilde{W}|$ and $\tilde{\mu}$ differ insignificantly from those given for $\tilde{\Delta}$ and $\tilde{\mu}$ in Table 9.1. This can be seen from Figs. 9.1 to 9.3, which give plots of $\tilde{\Delta}$, $\tilde{\mu}$, and ρ, respectively, obtained via. Eq. (9.8) and plots of $|\tilde{W}|$, $\tilde{\mu}$, and ρ

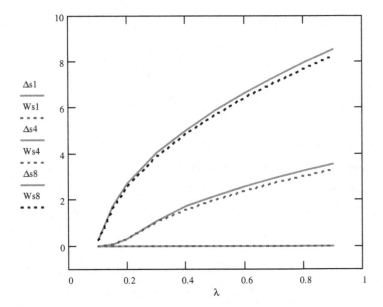

Fig. 9.1.   Variation of $\tilde{\Delta}$ (solid lines, obtained via Eqs. (9.8) and $|\tilde{W}|$ (dotted lines, obtained via the alternative equations discussed in the text) with λ for $E_F = E_{F1}$ (lower plot), E_{F2} (middle plot), and E_{F3} (upper plot).

Note: $E_{F1} = 10.01$, $E_{F2} = 1.01 \times 10^{-1}$, $E_{F3} = 2.17 \times 10^{-4}$ (eV).

obtained via Eq. (9.22) and the corresponding number equations for three values of n or E_F.

9.6.   Physical Significance of R and Variation of ρ as n is Decreased

The whole point of the exercise reported in the previous section is to establish that $\Delta \simeq |W|$, *even when chemical potential is retained in their respective equations*. Hence, Eq. (9.16) implies that Δ is also a measure of the change in the total energy of two electrons when they form a paired state. We are now enabled to interpret R defined in Eq. (9.12). Since Eq. (9.15) was derived in the rest frame of the pair in Chapter 3, values of Δ, $R > 1$ cannot be attributed to the motion of the c.m of the pair. Since $2E_F$ is the energy of two free electrons and $(2\mu + \Delta)$ the energy in their paired state, values of $R > 1$ signify a macroscopic occupation of a 2-electron state, i.e., conversion of the latter into a condensate.

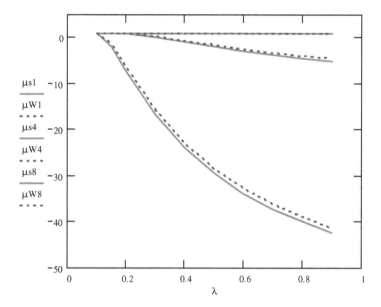

Fig. 9.2. Variation of $\tilde{\mu}$ with λ for $E_F = E_{F1}$ (upper plot), E_{F2}, (middle plot), and E_{F3} (lower plot) obtained via the set of Eqs. (9.8). Solid and dotted lines are to be distinguished as in the caption for Fig. 9.1.

Note: $E_{F1} = 10.01$, $E_{F2} = 1.01 \times 10^{-1}$, $E_{F3} = 2.17 \times 10^{-4}$ (eV).

Since we are using V as a handle, we can also envisage systems that are characterized by the same value of λ but different values of n or E_F. We have addressed such a scenario in Figs. 9.4, 9.5, and 9.6 which show, respectively, how $\tilde{\Delta}$, $\tilde{\mu}$, and R vary with E_F for $\lambda = 0.1$ and 0.9. Among these the plot for R is particularly instructive since it explicitly shows that that there are critical values of n or E_F *and* λ at which electron-pairs condense. Using Eq. (9.12) and Table 9.1, we find that

$$R(n = 1.448 \times 10^{14} \, \text{cm}^{-3}, \ \lambda = 0.9) = 1395. \qquad (9.24)$$

It seems interesting to observe that this number is similar[14] to the number of *molecules* in the (fermionic) condensate of ^{40}K when its atomic number density has nearly the same value as in Eq. (9.24). Hence CPs ought not to be regarded as automatically BE-condensed because there are critical values of n or E_F (and λ) given on the R versus n plots by the points beyond which R assumes values greater than unity: these describe macroscopic occupation of the paired state and belong to the realm of BEC.

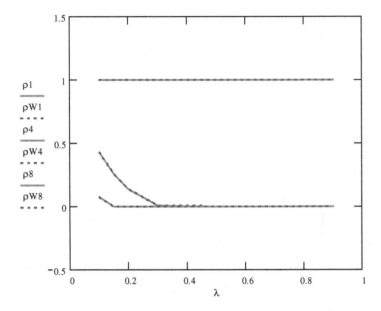

Fig. 9.3.   Variation of ρ with λ for $E_F = E_{F1}$ (upper plot), E_{F2}, (middle plot), and E_{F3} (lower plot). Solid and dotted lines are to be distinguished as in the caption for Fig. 9.1. *Note*: $E_{F1} = 10.01$, $E_{F2} = 1.01 \times 10^{-1}$, $E_{F3} = 2.17 \times 10^{-4}$ (eV).

The values of ρ in Table 9.1 shed light on the diminishing role of the hole–hole scatterings *vis-à-vis* electron–electron scatterings as one covers the spectrum of states from the extreme of the BCS limit on the left to the extreme of the BEC limit on the right.

9.7. Remarks

1. In Bogoliubov's treatment of the BCS theory, the upper limit[15] for the strength parameter λ has been set at 0.5. While we need to go beyond this limit to address the crossover picture, λ must remain less than unity because the equation that determines $|\tilde{W}|$ is obtained by summing a geometric series as was shown in Chapter 2. This accounts for the fact that we restricted ourselves to values of λ up to 0.9.

2. We note that the ratio ω_c/E_F increases as E_F is decreased for a fixed value of ω_c, this being so whether one solves the scaled Eq. (9.8), or the original Eq. (9.5). We note that *all* solutions given in Table 9.1 can also

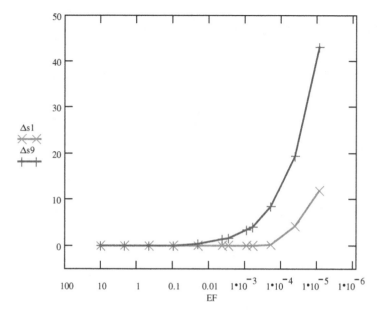

Fig. 9.4. Variation of $\tilde{\Delta}$ with *decreasing* $\ln(E_F)$ for $\lambda = 0.1$ (lower plot) and 0.9 (upper plot).

be obtained via the latter equations, procedural details for them being identical with those mentioned in Sec. 9.4. We chose to deal with the scaled equations in order to conform to the customary approach.

3. Indeed, for a given material and concentration, V (and hence λ) will have a specific value. So, it is prudent to ask: What is the significance of considering different values of λ for a given value of n or E_F? This begs the question: What causes the variation of the scattering length in the conventional treatment of the crossover? The answer to the latter question is: A change in the characteristics of the target and/or the particle/s scattered off it. Recall also that in experiments on trapped cold atomic gases which led to realization of BE condensates, scattering lengths and other parameters were tuned via the phenomenon of Feshbach resonance by adjusting an external parameter such as the magnetic field.[16] Consideration of a system characterized by different values of scattering length is equivalent to consideration of different systems. Returning now to the first question, consideration of different values of λ for a given value of n signifies dealing with samples that have different values of the BCS

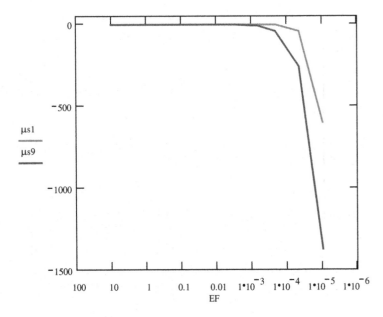

Fig. 9.5.   Variation of $\tilde{\mu}$ with *decreasing* ln(E_F) for $\lambda = 0.1$ (upper plot) and 0.9 (lower plot).

interaction parameter V; the point being that should one have such systems, one will find $\tilde{\Delta}$, etc., to vary in a continuous manner between the two regimes as is seen from Figs. 1–3.

4. We have shown in detail that the essential features of the crossover physics, i.e., continuity of the plots of $\tilde{\Delta}$ and $\tilde{\mu}$ as E_F is decreased, can be addressed without appealing to SLT. We also recall that there are infinitely many different shapes, depths and ranges of potentials that will reproduce a single scattering length, which is suggestive of why SLT does not shed light on the role of features peculiar to the BCS model interaction and embodied in the parameters ρ and R introduced in this chapter.

5. We believe that the intuitive approach to crossover problem followed above will make the subject more accessible to readers unfamiliar with the intricacies of regularization or renormalization that the customary approach requires. An important outcome of this chapter is that it opens up new avenues for dealing with real life SCs such as $SrTiO_3$ because we have learnt to deal with the equations for Δ and μ — without appealing

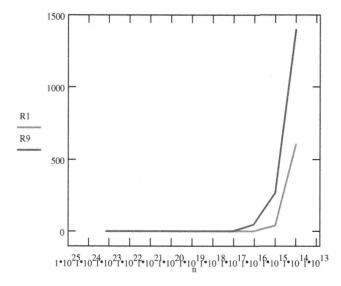

Fig. 9.6.   Variation of R with *decreasing* ln(n) for λ = 0.1 (lower plot) and 0.9 (upper plot).

to SLT or the need to introduce the parameter A defined vide Eq. 7 in Eagles' paper.[1] We take up in the next chapter a study of this SC which has been the cause of a puzzle that has persisted for nearly 50 years.

Notes and References

1. D.M. Eagles, *Phys. Rev.* **186**, 456 (1969).
2. The existence of pre-formed Cooper pairs before superconductivity sets in has been suggested as a possible explanation for the pseudo gaps that have been observed in some SCs. See, for example, T.A. Mamedov and M. de Llano, *J. Supercond. Nov. Magn.* **28**, 291 (2015) and references therein.
3. I. Bardeen, L.N. Cooper and J.R. Schrieffer, *Phys. Rev.* **108**, 1175 (1957).
4. For a review, see: J.M. Blatt, *Theory of Superconductivity* (Academic Press, New York, 1964).
5. A.J. Leggett, in: *Modern Trends in the theory of Condensed Matter* (Springer-Verlag, Berlin, 1980) p. 13; J. Phys. (Paris) **41**, C7–C19 (1980).
6. M. Marini, F. Pistolesi, and G.C. Strinati, *Eur. Phys. J. B* **1**, 151(1998).
7. T. Papenbrock and G.F. Bertsch, *Phys. Rev. C* **59**, 2052 (1999).
8. R.M. Quick, C. Eseberg and M. de Llano, *Phys. Rev. B* **47**, 11512 (1993); Carter *et al.*, *Phys. Rev. B* **52**, 16149 (1995).
9. P. Nozières and S. Schmitt-Rink, *JLTP* **59**, 195 (1985).
10. Q. Chen *et al.*, *Phys. Reps.* **412**, 1 (2005).
11. M. Randeria, in: *Bose-Einstein Condensation*, eds., A. Griffin, D.W. Snoke and S. Stringari (Cambridge University Press, 1995), p. 355, and references therein.

12. For a popular account, further references, and pictorial representation of the BE-condensates, see, for example, K. Burnett, M. Edwards and C.W. Clark, *Physics Today*, Dec. 1999, p. 37.

13. D.M. Eagles, *Phys. Rev.* **178**, 668 (1969).

14. C. Regal, *PhD thesis* (The University of Colorado, Boulder, 2005).

15. Ref. 4, p. 206.

16. C.J. Pethick and H. Smith, *Bose-Einstein Condensation in Dilute Gases* (Cambridge University Press, Cambridge, 2002).

This chapter is based on: G.P. Malik, *Int. J. Mod. Phys. B* **28**, 1450054 (2014) (13 pages).

Chapter 10

On the Puzzle Posed by Superconducting $SrTiO_3$

10.1. Introduction

10.1.1. *The nature of the puzzle*

While ordinarily an insulator, upon doping with n-type carriers, $SrTiO_3$ was found[1] in 1964 to undergo a superconducting transition below 1 K. A remarkable feature of $SrTiO_3$ in this state is that its T_c is a function of the electron concentration n: For $6.9 \times 10^{18} \leq n \leq 5.5 \times 10^{20}$ cm^{-3}, the plot of T_c roughly resembles a *dome* the apex of which corresponding to $n \approx 9 \times 10^{19}$ cm^{-3} has the value of $T_c \approx 0.3$ K. On *either* side of this value, T_c goes down to ≈ 0.1 K for the two extreme values between which n was varied.[2] Note that this implies that, experimentally, one obtains the same value of T_c for two different values of n. The puzzle for the theory has been to explain this result; it is addressed in this chapter on the basis of our treatment of the μ-incorporated equations given in the last chapter.

10.1.2. *Earliest work on the puzzle*

While presenting the experimental results on the variation of T_c as a function of n, Koonce *et al.*[2] also offered a theoretical explanation in terms of the normal-state properties of $SrTiO_3$ by introducing some novel ideas. Among these was the postulate that the kernel of the T-dependent BCS gap equation should have three terms: One due to intravalley Coulomb and phonon interactions, and the other two due to the intervalley Coulomb and the intervalley phonon interactions, respectively. In a formidable piece of theory, they employed well-reasoned values of 19 parameters — such as the renormalized effective mass, band effective mass, number of valleys in the conduction band, etc. — to calculate the transition temperatures. Further, since the deformation potential is not known for $SrTiO_3$, they chose a value for it that led to a T_c in agreement with the experimental value at a given value of n. This value of the intervalley deformation potential was then used to calculate T_c at all other values of n. In this manner they obtained a one-parameter fit to the observed data. However, they also noted that "it was quite difficult to obtain even a crude fit to the curve of transition temperatures as a function of concentration using normal-state values noticeably different from those given in (their) Table 10.1."

In a serious follow up of the above work, Eagles[3] addressed the problem by additionally taking into account the data on the magnetic field penetration depths in Zr-doped $SrTiO_3$. From these he deduced that $m_3/m_1 \simeq 1.48$, where m_1 (m_3) denotes the effective mass when 1% (3%) of the Ti ions are replaced by Zr ions which are regarded as the main scattering centres. His model — a simpler version of the one employed in Ref. 2 — involved screened electron–electron interactions due to intervalley phonons of energy

Table 10.1. Solutions obtained for LHS of the dome ($2.5 \leq \mu < 15$ meV). Values of $n(\mu)$ are obtained via Eq. (10.12). The values of λ in the first and the last rows are obtained via *truncated* form of Eq. (10.14) — for which $I_3(\mu, T_c)$ is as given in Eq. (10.23) — with the input of (μ, T_c) values in the same rows. The value of λ in each of the remaining rows is obtained via Eq. (10.24); the input of μ and λ in the row is then used to calculate T_c via the truncated form of Eq. (10.14). The first two numbers in the last column give the TOLERANCE (as the exponent of 10) with which $I_3(\mu, T_c)$, and $I_4(\mu, T_c)$, respectively, were calculated, and the last number gives the TOL with which the equation under consideration was then solved.

μ (meV)	$n(\mu)$ (cm^{-3})	λ	T_c (K)	TOL
2.5	6.73×10^{18}	0.12844	0.07	$-6.5, -8, -8$
5.0	1.90×10^{19}	0.13402	0.087	$-6, -8, -8$
7.5	3.50×10^{19}	0.13960	0.121	$-6.5, -8, -8$
10.0	5.39×10^{19}	0.14518	0.167	$-7.5, -8, -8$
12.5	7.53×10^{19}	0.15076	0.228	$-7.5, -8, -8$
14.9	9.79×10^{19}	0.15611	0.3	$-7.5, -8, -8$

0.0497 eV modified by intervalley Coulomb repulsion and by intravalley effects. Postulating further that the ratio of phonon-induced to Coulomb intervalley interactions should increase by about 5% for each percent of Zr doping and by employing four arbitrary parameters in the model, he obtained a rather good fit for the T_c versus n curve.

10.1.3. *Recent findings about SrTiO₃*

That the puzzle persists to this day is evidenced by the statements made in 2013 by Lin *et al.*[4] that "The origin of superconductivity in bulk SrTiO₃ is a mystery ... ," and "The microscopic mechanism of pair formation in bulk SrTiO₃ remains a wide-open question." While not attempting the ambitious task of theoretically obtaining the T_c versus n curve, the latter papers addressing the puzzle have nevertheless shed light on new important features of SrTiO₃. We now give a brief account of a few of them which are relevant to our approach. Among these are the papers by: (i) Binnig *et al.*[5] which, based on tunnelling measurements, reported that SrTiO₃ has a distinctive two-band structure and is characterized by two well-defined energy gaps; (ii) Bussmann-Holder *et al.*[6] who carried out a detailed analysis of SrTiO₃ in a two-band model of superconductivity and reported, among other findings, that SrTiO₃ should exhibit a point of inflection close to T_c and not close to $T = 0$; (iii) Lin *et al.*[4] who, employing the Nernst effect which is an extremely sensitive probe of tiny bulk Fermi surfaces, reported that SrTiO₃ has both a sharp Fermi surface and a superconducting ground state down to concentrations as low as 5.5×10^{17} cm^{-3}; (iv) Jourdan *et al.*[7] who, regarding the doped SrTiO₃ as a candidate for bipolaronic superconductivity, concluded via their analysis that its "special superconducting properties can be related to its Fermi surface." Finally, it is noteworthy that in a more recent paper, Lin *et al.*[8] have reported that the T_c versus n plot for SrTiO₃ is characterized by *two* domes — this is a feature that we shall briefly address below.

10.2. Dealing with SrTiO₃ via Equations Incorporating Chemical Potential

Since the salient feature of SrTiO₃ is variation in the values of its T_c and Δ_0 as n (or μ) is varied, it is natural to ask: why not incorporate μ into the equations for T_c and Δ_0 and then regard it as an independent variable? Obtained in the previous chapter, such equations are applied here to SrTiO$_3$.

Our starting point is provided by another paper by Eagles.[9] In this paper the BCS gap equation was studied in conjunction with the equation for n in superconducting semiconductors at $T = 0$ under the assumption of large cut-off energies: $k_B\theta_D/E_F \gg 1$, i.e., for low values of n. We recall that E_F does not figure in the BCS equation for Δ as it is usually written because it is assumed that

$$k_B\theta_D/E_F \ll 1 \qquad (10.1)$$

and E_F is taken as the zero of the energy scale. In dealing with SrTiO$_3$ we need to do away with constraint Eq. (10.1). Such a step was shown in the last chapter to lead to Eqs. (9.5). These equations are rewritten below by replacing ω_c by $k_B\theta_D$ and by making the temperature-dependence of the variables explicit: $E_F(0)$, λ_0, μ_0, and Δ_0 denote their values at $T = 0$.

$$[E_F(0)]^{1/2} = \frac{\lambda_0}{2} I_1(\mu_0, \Delta_0),$$

$$I_1(\mu_0, \Delta_0) = \int_{-k_B\theta_D}^{k_B\theta_D} d\xi \frac{\sqrt{\xi + \mu_0}}{\sqrt{\xi^2 + \Delta_0^2}} \qquad (10.2)$$

$$\frac{4}{3}[E_F(0)]^{3/2} = I_2(\mu_0, \Delta_0),$$

$$I_2(\mu_0, \Delta_0) = \int_{-\mu_0}^{k_B\theta_D} d\xi \sqrt{\xi + \mu_0}\left[1 - \frac{\xi}{\sqrt{\xi^2 + \Delta_0^2}}\right],$$

$$= \frac{4}{3}(\mu_0 - k_B\theta_D)^{3/2}$$

$$+ \int_{-k_B\theta_D}^{k_B\theta_D} d\xi \sqrt{\xi + \mu_0}\left[1 - \frac{\xi}{\sqrt{\xi^2 + \Delta_0^2}}\right], \qquad (10.3)$$

since $\Delta_0 = 0$ for $-\mu \leq \xi \leq -k_B\theta_D$: in this region the expression in the square brackets in the upper equation for I_2 (μ_0, Δ_0) reduces to 2. Further, we have

$$\lambda_0 = \left[\frac{(2m/\hbar^2)^{3/2} E_F(0)^{1/2}}{4\pi^2} \right] V_0, \qquad (10.4)$$

and V_0 is a term due to both Coulomb repulsion between a pair of electrons and attraction between them because of the ion-lattice at $T = 0$.

From Eqs. (10.2) and (10.3) we obtain

$$\frac{\lambda_0}{2} I_1(\mu_0, \Delta_0) - \left[\frac{3}{4} I_2(\mu_0, \Delta_0) \right]^{1/3} = 0, \qquad (10.5)$$

which is one of our basic equations. After μ_0 and Δ_0 are determined, we can calculate $E_F(0)$ via

$$E_F(0) = \left[\frac{3}{4} I_2(\mu_0, \Delta_0) \right]^{2/3}. \qquad (10.6)$$

We now consider the equation for T_c (Eq. (3.24) in Chapter 3, with E_F replaced by μ_1 and $W = 0$):

$$1 = \frac{V_1}{(2\pi)^3} \frac{1}{2} \int_{\mu_1 - k_B\theta_D}^{\mu_1 + k_B\theta_D} d^3 p \frac{\tanh[\frac{\beta_c}{2}(p^2/2m - \mu_1)]}{(p^2/2m - \mu_1)}, \qquad (\beta_c = 1/kT_c) \qquad (10.7)$$

where V_1 and μ_1 denote their values at $T = T_c$.

With $\xi = p^2/2m - \mu_1$, Eq. (10.7) becomes

$$1 = \frac{(2m)^{3/2} V_1}{4\pi^2} \int_{-k_B\theta_D}^{k_B\theta_D} d\xi \frac{\sqrt{\xi + \mu_1} \tanh(\xi/2kT_c)}{2\xi}. \qquad (10.8)$$

Note that if were to replace μ_1 by E_F in this equation and assume Eq. (10.1), then we would obtain the equation for T_c in the more familiar form. On the other hand if we multiply Eq. (10.8) with $[E_F(T_c)]^{1/2}$ we obtain

$$[E_F(T_c)]^{1/2} = \frac{\lambda_1}{2} I_3(\mu_1, T_c),$$

$$I_3(\mu_1, T_c) = \int_{-k_B\theta_D}^{k_B\theta_D} d\xi \frac{\sqrt{\xi + \mu_1} \tanh(\xi/2kT_c)}{\xi}, \qquad (10.9)$$

where we have put

$$\frac{(2m/\hbar^2)^{3/2} V_1 E_F^{1/2}(T_c)}{4\pi^2} = \left[\frac{(2m/\hbar^2)^{3/2} V_0 E_F^{1/2}(0)}{4\pi^2}\right] \frac{E_F^{1/2}(T_c)}{E_F^{1/2}(0)} \frac{V_1}{V_0}$$

$$= \lambda_0 \sqrt{\frac{\mu_1}{\mu_0}} \frac{V_1}{V_0} \equiv \lambda_1, \tag{10.10}$$

used Eq. (10.4), assumed that $E_F(0) = \mu(0)$ and $E_F(T_c) = \mu_1$ — to be justified *a posteriori*.

We now need the number equation at $T = T_c$ which is obtained from[10]:

$$n(\beta) = \int_0^{k_B \theta_D + \mu} \frac{d^3 p}{(2\pi)^3}$$

$$\times \left[1 - \frac{(p^2/2m - \mu) \tanh\left(\frac{\beta}{2}\left(\sqrt{(p^2/2m - \mu)^2 + \Delta^2}\right)\right)}{\sqrt{(p^2/2m - \mu)^2 + \Delta^2}}\right]. \tag{10.11}$$

Note that when $\beta = \infty$ (i.e., $T = 0$), we have $\mu = \mu_0$, $\tanh(\ ) = 1$, and $\Delta = \Delta_0$, whence $\xi = p^2/2m - \mu_0$ leads to Eq. (10.3) above. On the other hand when $T = T_c$, we have $\mu = \mu_1$ and $\Delta = 0$; then using

$$n(T_c) = \left[\frac{1}{3\pi^2} \{2m E_F(T_c)/\hbar^2\}^{3/2}\right], \quad \xi = p^2/2m - \mu_1, \tag{10.12}$$

we have Eq. (10.11) as

$$\frac{4}{3}[E_F(T_c)]^{3/2} = I_4(\mu_1, T_c),$$

$$I_4(\mu_1, T_c) = \int_{-\mu_1}^{k_B \theta_D} d\xi \sqrt{\xi + \mu_1} \left[1 - \tanh\left(\frac{\xi}{2k_B T_c}\right)\right]. \tag{10.13}$$

Eliminating $E_F(T_c)$ from Eqs. (10.9) and (10.13), we obtain

$$\frac{\lambda_1}{2} I_3(\mu_1, T_c) - \left[\frac{3}{4} I_4(\mu_1, T_c)\right]^{1/3} = 0, \tag{10.14}$$

Eq. (10.14) is our second basic equation. Equations (10.5) and (10.14) are to be solved subject to the relation between λ_0 and λ_1 given in Eq. (10.10) i.e.,

$$\lambda_1 = \lambda_0 \sqrt{\frac{\mu_1}{\mu_0}} \frac{V_1}{V_0}.$$

If λ_1, μ_1, and T_c can be determined, then $E_F(T_c)$ follows from

$$E_F(T_c) = \left[\frac{\lambda_1}{2} I_3(\mu_1, T_c) \right]^2. \qquad (10.15)$$

In the next section we deal with how Eqs. (10.5), (10.10), and (10.14) provide a direction for the solution of the puzzle posed by SrTiO₃.

10.3. Solutions of Eqs. (10.5) and (10.14) as μ is Varied

10.3.1. *The Debye temperature of Ti ions*

In Ref. 2 or 3 no appeal is made to θ_D of SrTiO₃. However, the input of this parameter is essential in our approach. The values available in the literature for θ_D of SrTiO₃ differ considerably — while Roth *et al.*[11] cite it as 490 K, the value quoted by Burfoot and Taylor[12] is 700 K. We adopt here the former of these values.

We now recall that Debye temperature is just another way to specify the Debye frequency and is not to be confused with thermodynamic temperature. Therefore $\theta_D(\text{SrTiO}_3) = 490$ K does *not* mean that θ_D of the Sr or the Ti ions — one phonon exchanges with either of them separately must be regarded as responsible for pairing and hence for superconductivity — is 490 K. Their θ_Ds *must* be different because the mass of an Sr ion is different from that of a Ti ion. However, since SrTiO₃ is a perovskite that contains a sub-lattice of layers of SrO and another sub-lattice of layers of TiO₂, one may assume that θ_D for each of these sub-lattices is 490 K. In this chapter we mainly consider the TiO₂ sub-lattice. The question now is: Given $\theta_D(\text{TiO}_2) = 490$ K, what are $\theta_D(\text{Ti})$ and $\theta_D(\text{O})$ (the latter of course plays no role in the pair-formation)? For an alloy/compound $A_x B_{1-x}$, the following relation has been routinely used in the literature[13]:

$$\theta_D(A_x B_{1-x}) = x\theta_D(A) + (1 - x)\theta_D(B). \qquad (10.16)$$

Identifying $A_x B_{1-x}$ with O_2Ti, we have the LHS of Eq. (10.16) as 490 K and $x = 2/3$, whence

$$490 = (2/3)\theta_D(O) + (1/3)\theta_D(Ti). \tag{10.16a}$$

To determine $\theta_D(O)$ and $\theta_D(Ti)$ we now require another equation. This requirement is met by assuming that the oscillations of the Ti and the O ions simulate the small oscillations of a coplanar double pendulum. Assuming that Ti is the *lower* bob of the double pendulum, we have

$$\frac{\theta_D(O)}{\theta_D(Ti)} = \left[\frac{1 + \sqrt{m_{Ti}/(m_{Ti} + m_O)}}{1 - \sqrt{m_{Ti}/(m_{Ti} + m_O)}} \right]^{1/2}, \tag{10.17}$$

where m_O (m_{Ti}) is the atomic mass number of an O (Ti) ion. A derivation of Eq. (10.17) was given in Chapter 4. With $m_O = 15.999$ and $m_{Ti} = 47.867$, the solution of Eqs. (10.16a) and (10.17) gives

$$\theta_D(Ti) = 173.9 \text{ K.} \quad \text{(and } \theta_D(O) = 648.1, \text{ which we do not require)} \tag{10.18}$$

This is the value of $\theta_D(Ti)$ adopted in this study.

10.3.2. *Fixing values of μ_1 for which Eqs. (10.5) and (10.14) need to be solved*

Since the task of the theory is to provide values of T_c corresponding to different values of n (equivalently, values of μ_1 at $T = T_c$), we now need to fix the values of μ_1, which we treat as an independent variable, in order to solve Eqs. (10.5) and (10.14). The relation between μ_1 and the number density $n(\mu_1)$ of the carriers is given by Eq. (10.12) with m and E_F replaced by m_d (the density-of-states mass) and μ_1, respectively. Following Koonce et al.,[2] we take

$$m_d = 3^{2/3} \times 2.5 \times m_e. \quad (m_e = \text{free electron mass}) \tag{10.19}$$

From Eqs. (10.12) and (10.19) we obtain

$$n(2.5 \text{ meV}) = 6.73 \times 10^{19} \text{ cm}^{-3}, n(15 \text{ meV}) = 19.89 \times 10^{19} \text{ cm}^{-3},$$
$$n(45 \text{ meV}) = 5.14 \times 10^{20} \text{ cm}^{-3} \tag{10.20}$$

We now note that:

(a) The range of carrier concentration n in the smooth dome-like dashed T_c versus n plot *"drawn for illustrative purposes"* in the data of Ref. 2 is $6.9 \times 10^{19} \leq n \leq 5.5 \times 10^{20}$ (cm^{-3}), which is very nearly covered by $2.5 \leq \mu_1 \leq 45$ (meV). Further, in this plot

(b) The maximum value of T_c given by the apex of the dome, i.e., ≈ 0.3 K, corresponds to the value of $n(15 \text{ meV})$ in Eq. (10.20).

(c) For the value of $\theta_D = 173.9$ fixed by us, $k_B \theta_D \simeq 14.99$ meV.

It thus follows that the values of T_c on the LHS of the dome correspond to $2.5 \leq \mu_1 < 15$ (meV) whereas those on the RHS of the dome correspond to $15 < \mu_1 \leq 45$ (meV). Because of this asymmetry, we undertake below the calculation of T_cs on the LHS of the dome at six values of μ_1 between 2.5 and 15 meV by varying it in steps of 2.5 meV, and at seven points between 15 and 45 meV on the RHS by varying it in steps of 5 meV.

10.3.3. *Equations (10.5), (10.10) and (10.14) comprise an under-determined system; approximate solutions for LHS of the dome (2.5 meV ≤ μ_1 < 15 meV) with suitable assumptions*

Having fixed the values of θ_D and μ_1 as above, it is seen that we have *five* unknowns, $\mu_0, \lambda_0, \lambda_1, \Delta_0$, and T_c, but *three* Eqs. (10.5), (10.10), and (10.14). This is so because our treatment of the problem so far has been perfectly general.

We must now make reasonable assumptions in order that the problem becomes tractable. To this end we note that we differentiated above between $E_F(T_c)$, or μ_1, and $E_F(0)$, or μ_0. The rationale for this is: While all the electrons in the system are free at $T = T_c$ (assuming that there are no pre-formed pairs for $T \geq T_c$), a vast majority of them ends up as Cooper pairs at $T = 0$. Recalling that chemical potential signifies the energy required to add a fermion to a many-fermion system, the decrease in the number of free electrons in going from T_c to $T = 0$ implies that (a) μ_1 must be *greater* than μ_0 and (b) V_1 must be *smaller* than V_0 — since fewer electrons at $T = 0$ cause a reduction in the repulsive part of the interaction. These considerations are reminiscent of the fact — dealt with in Chapter 5 — that thermal conductivity (akin to λ here) of an SC is not as adversely affected by the formation of pairs (i.e., by the decrease in the number of heat carriers) as one would expect because this effect is offset by an increase in the mean free path of the electrons. Hence we now assume that

$$\mu_0 = \mu_1 \equiv \mu, \quad V_0 = V_1, \quad \text{whence via Eq. (10.10), } \lambda_1 = \lambda_0 \equiv \lambda. \tag{10.21}$$

The equality in the last of the above equations is *factually* a basic tenet of the BCS theory. We could have assumed it at outset, but did not because we are dealing here with a puzzle that has defied solution for nearly five decades. This warrants that all aspects of the problem that may have a bearing on the solution should be kept in view.

As a consequence of the above considerations, with μ as an independent variable, we are left with two Eqs. (10.5) and (10.14), but three unknowns: λ, Δ_0, and T_c. However, with the input of different assumed values of λ corresponding to, say, $\mu = 2.5$ meV — which corresponds to the lowest value of n in the data of Ref. 2 — we can solve these two equations and find the value of λ that leads to $T_c \approx 0.1$ K. Attempting to carry out this first step of our calculations, we find that $I_1(\mu, \Delta_0)$, $I_2(\mu, \Delta_0)$ and $I_3(\mu, T_c)$ become imaginary. This happens because the lower limit in the integrals for all of these is $\xi = -k_B \theta_D = -14.99$ meV and we have the factor $\sqrt{\xi + \mu}$ in their integrands. This is a problem that we have already encountered in the study of the BCS–BEC crossover physics in the last chapter. In order to obtain real solutions of our problem, we must now set the lower limit in each of the three integrals at $-\mu$ — as was also done in the crossover study. On doing so we have

$$I_1^t(\mu, \Delta_0) = \int_{-\mu}^{k_B \theta_D} d\xi \, \frac{\sqrt{\xi + \mu}}{\sqrt{\xi^2 + \Delta_0^2}},$$

$$I_2^t(\mu, \Delta_0) = \int_{-\mu}^{k_B \theta_D} d\xi \sqrt{\xi + \mu} \left[1 - \frac{\xi}{\sqrt{\xi^2 + \Delta_0^2}} \right], \tag{10.22}$$

$$I_3^t(\mu, T_c) = \int_{-\mu}^{k_B \theta_D} d\xi \, \frac{\sqrt{\xi + \mu} \, \tanh(\xi/2kT_c)}{\xi}, \tag{10.23}$$

where the superscript t denotes *truncated*. Note that the term $(4/3)(\mu - k_B\theta_D)^{3/2}$ in Eq. (10.3) for $I_2(\mu, \Delta_0)$ is no longer present in the second of the two equations in Eqs. (10.22).

Replacing $I_1(\mu, \Delta_0)$, $I_2(\mu, \Delta_0)$ and $I_3(\mu, T_c)$ in Eqs. (10.5) and (10.14) by the expressions given above, we now proceed as follows.

(A) Solve Eq. (10.14) with the input of (a) $\mu = 2.5$ meV and $T_c = 0.07$ K to obtain $\lambda = 0.12844$ and (b) $\mu = 14.9$ meV and $T_c = 0.3$ K to obtain $\lambda = 0.15611$.

(B) Assuming a linear variation of λ between the above values of μ, we have

$$\lambda(\mu) = 0.12844 + 2.2315(\mu - 2.5 \times 10^{-3}).$$

$$(2.5 \text{ meV} \leq \mu < 15 \text{ meV}) \qquad (10.24)$$

Using Eq. (10.24), we obtain the values of λ corresponding to $\mu = 5$, 7.5, 10, and 12.5 (meV).

(C) For each pair of (μ, λ) values in (B), we solve Eq. (10.14) to obtain the corresponding value of T_c. Equation (10.15) is then used to calculate E_F.

In principle, for each pair of (μ, λ) values in (A) and (B) one can solve Eq. (10.5) to obtain the corresponding value of Δ_0. We give in this chapter the value corresponding to the apex of the dome only, where it can be calculated with a TOLERANCE (TOL) of $10^{-7.5}$. Farther from the apex such solutions can generally be found only with a TOL of $\approx 10^{-5}$, or worse. Since Eq. (10.5) involves Δ_0^2 which is of the order of $\approx 10^{-10}$ eV, these solutions cannot be trusted — they ought to be calculated by employing quadruple precision.

The results of the above calculations have been given in Table 10.1.

10.3.4. Approximate solutions for RHS of the dome (15 meV < μ ≤ 45 meV)

Since $\mu_1 > k_B\theta_D$ on the RHS of the dome, we no longer need to curtail the limits of the integrals in Eqs. (10.5) and (10.14). Using Eq. (10.21), the procedure now is similar to the one followed for LHS of the dome. Solution of Eq. (10.15) with the input of $T_c = 0.07$ K and $\mu = 15.1$ (45) meV now gives $\lambda = 0.15602$ (0.126). Therefore the equation corresponding to Eq.

Table 10.2.   Solutions obtained for RHS of the dome ($15 < \mu \leq 45$ meV). Values of $n(\mu)$ are obtained via Eq. (10.12). The values of λ in the first and the last rows are obtained via Eq. (10.14) with the input of (μ, T_C) values in the same rows. The value of λ in each of the remaining rows is obtained via Eq. (10.25); the input of μ and λ in the row is then used to calculate T_C via Eq. (10.14). See the caption for Table 10.1 for significance of the numbers in the last column.

μ (meV)	$n(\mu)$ (cm^{-3})	λ	T_c (K)	TOL
15.1	9.99×10^{19}	0.15602	0.3	$-7.5, -8, -8$
20.0	1.52×10^{20}	0.15110	0.2533	$-7.5, -8, -8$
25.0	2.13×10^{20}	0.14608	0.2049	$-7.5, -8, -8$
30.0	2.80×10^{20}	0.14106	0.1618	$-7.5, -8, -8$
35.0	3.53×10^{20}	0.13604	0.1249	$-6.5, -8, -8$
40.0	4.31×10^{20}	0.13102	0.0946	$-6, -7.5, -8$
45.0	5.14×10^{20}	0.126	0.07	$-6, -7, -8$

(10.24) is

$$\lambda(\mu) = 0.15602 - 1.00435(\mu - 15.1 \times 10^{-3}). \quad (15 \text{ meV} < \mu \leq 45 \text{ meV})$$
$$(10.25)$$

Carrying out the remaining steps as noted for LHS of the dome, the results of the calculations have been given in Table 10.2.

Since the value of E_F corresponding to the pair of (λ, μ) values in any row in both the tables was found to be equal (up to four places of decimals) to the value of μ in the same row, its values are not quoted there. This result provides the *a posteriori* justification mentioned above, preceding Eq. (10.12).

Fig. 10.1 gives the T_c vs. $\ln(n)$ plot for both sides of the dome.

10.4.   A Purely Mathematical Model to Address the T_c versus n plot in Ref. 2

When a puzzle remains unsolved for as long as the one being addressed here, it is prudent to consider new ideas even though their physical basis might not be immediately evident. It is in this spirit that we offer below a purely mathematical model for λ as a function of the reduced chemical

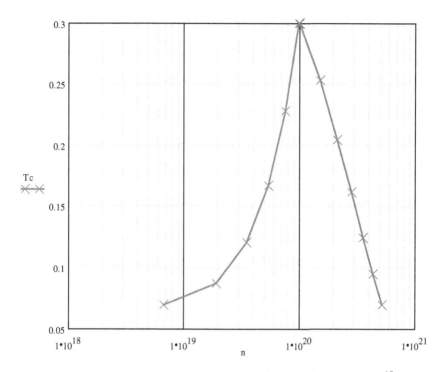

Fig. 10.1. Plot of T_c vs. ln(n) obtained via solutions of (15) for $6.73 \times 10^{18} \leq n \leq 5.14^{20}$ cm^{-3}. Solutions to the left of the apex are obtained by curtailing the lower limits of $I_1(\mu, \Delta_0)$, $I_2(\mu, \Delta_0)$, and $I_3(\mu, T_c)$ as discussed in the text.

potential μ_r:

$$\lambda(\mu_r) = \frac{\lambda_a}{2}\left[1 + \frac{1}{1 + g(\mu_r - 1)^2}\right], \qquad (10.26)$$

where $\mu_r = \mu/\mu_a$, μ_a being the value of the chemical potential corresponding to the apex of the dome, λ_a the value of the interaction parameter obtained by solving Eq. (10.15) with the input of μ_a and the corresponding value of T_c, and g as constant. For the data of Ref. 2, we have

$$2.5 \leq \mu \leq 45(\text{meV}), \quad \mu_a = 15 \text{ meV}, \quad T_c = 0.3 \text{ K.} \qquad (10.27)$$

With

$$g = 0.5 \quad \text{for } \mu < 15 \text{ meV}, \quad \text{and} \quad g = 0.15 \quad \text{for } \mu > 15 \text{ meV} \qquad (10.28)$$

we calculate via Eq. (10.26) the values of λ corresponding to the six points on the LHS of the dome and seven points on the RHS of the dome that we considered above. With the input of these (μ, λ) values into Eq. (10.15) and (10.55) – employing Eqs. (10.22) and (10.23) on the LHS — we can obtain the values of T_c and Δ_0 for our model. These results have been given in Table 10.3 and the corresponding T_c versus μ plot in Fig. 10.2.

10.5. On the Feature of Two Domes in the T_c versus n Plot for SrTiO₃ Reported by Lin *et al.*[8]

It was noted above that Lin *et al.*[8] have recently reported that the T_c versus n plot for SrTiO₃ is characterized by *two* domes, with ($n \approx 10^{18}$ cm^{-3}, $T_c \approx 0.2$ K) as the apex of the lower dome and ($n \approx 10^{20}$ cm^{-3}, $T_c \approx 0.4$ K) that of the upper dome. In this context we note that we have so far confined ourselves to the formation of Cooper pairs via phonon interactions with the Ti ions. If we additionally take into account formation of pairs via interaction with the Sr ions, we are led via Eqs. (10.16) and (10.17), to $\theta_D(\text{Sr}) = 166.5$ K. Hence it is suggested that interactions with the Ti ions and the Sr ions, separately, at different values of n, are the likely cause of the two domes observed in SrTiO₃. The origin of the two gaps in SrTiO₃ can also be understood via the difference in the binding energies of the pairs thus formed, recalling that (half) the binding energy of a pair equals Δ.

10.6. Remarks

1. Equations (10.2) and (10.3) considered as a couple of simultaneous equations have been employed in a large number of papers in the context of the BCS–BEC crossover physics — usually in the framework of the scattering length theory. Obtaining Eq. (10.5) from these equations was the first crucial, *new*, step in our study. Combining together Eqs. (10.9) and (10.13) to obtain Eq. (10.14) was a second similar step.

2. We chose in our study to differentiate between λ_0 and λ_1 — despite it being a basic tenet of the BCS theory that the two should be equal — because we found in Chapter 3 that, empirically, value of the former calculated via the equation for Δ_0 is *invariably* different from that of

Table 10.3. Values of λ as a function of μ obtained via the model in Eq. (10.26). With the input of these (μ, λ) values, the T_c values are calculated via Eq. (10.14) — using truncated expressions given in Eq. (10.22) for $\mu < 15$ meV. The value of Δ_0 corresponding to $\mu = 15$ eV calculated via Eq. (10.5) is 4.524×10^{-5} eV and that of $2\Delta_0/k_B T_c$ is 3.51. Values of Δ_0 for other values of μ are not calculated for the reason given in the text.

μ (meV)	2.5	5	7.5	10	12.5	15	20	25	30	35	40	45
λ	0.1359	0.1418	0.1473	0.1519	0.1549	0.156	0.1551	0.1527	0.1489	0.1442	0.139	0.1337
T_c (K)	0.107	0.131	0.176	0.227	0.272	0.300	0.300	0.276	0.235	0.190	0.147	0.110

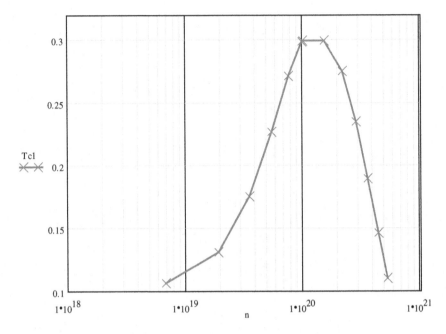

Fig. 10.2.   Plot of T_c vs. ln(n) for the model given in Eq. (10.26) with the same specifications as for Fig. 10.1.

the latter calculated via the equation for T_c. Relating λ_0 and λ_1 via Eq. (10.10) formed the third step of our study.

In essence the approach followed in this chapter parallels the approach that is followed in the BCS theory for dealing with the elemental SCs, where λ is calculated via the equation for T_c (or Δ_0) and then used to predict Δ_0 (or T_c). In order to go beyond the approximations that we made in Eq. (10.21), we need *precise* experimental values of additional variables, such as Δ_0 at μ_0 because the equations now involve additional variables. Another approach to supplement Eq. (10.5) and (10.14) could be to adapt the work of Van Marel and collaborators[14,15] who have studied variation of μ with n, and T, in different contexts.

3. Our study in this chapter has shed light on the most important feature of the puzzle posed by the T_c versus μ plot given in Ref. 2: Increase in T_c with μ on the LHS of the apex, but decrease on the RHS of it. We have shown this to be a logical consequence of the need to use "truncated" equations on the LHS of the dome. We recall that such a need

also arises for the same reason (i.e., to avoid imaginary solutions) in a study[16] of various Bose condensates via a linearly perturbed harmonic oscillator potential, as also in a study of the BCS–BEC crossover physics as was seen in the last chapter. This nonetheless raises the question: Is there a *physical* explanation for truncation? We note in this connection that our simplified treatment of the problem is based on an *effective* mean-field description of conduction electrons in a single band, whereas SrTiO₃ is known to be characterized by multiple bands. This feature and itinerancy of the electrons suggest that the need for truncation arises because electrons can be, as it were, in a "rain shadow" region where the entire spectrum of phonon frequencies is not available for the formation of pairs.

4. Because the puzzle addressed here has remained unsolved for a long time, it seems to us that one should not be averse to considering new ideas even though there physical basis is not immediately evident. It is in this spirit that a purely mathematical for $\lambda(\mu_r)$ was also given above. While presumably the model can be further refined — by choosing different values for g in Eq. (10.21) for two sides of the dome, it would be interesting to find a physical justification for it.

5. It also seems interesting to ask: What would be the T_c of SrTiO₃ if, through ingenuity, one could cause pairing via simultaneous exchanges of phonons with both the Ti and the Sr ions? A rough estimate of T_c in such a situation may be obtained by invoking the following equation which incorporates two-phonon exchange mechanism and was dealt with in Chapter 4:

$$1 = \lambda_{Ti} \int_0^{\theta_D(Ti)/2T_c} dx \frac{\tanh(x)}{x} + \lambda_{Sr} \int_0^{\theta_D(Sr)/2T_c} dx \frac{\tanh(x)}{x}$$

For values of $\theta_D(\text{Ti}) = 173.9$ K, $\lambda_{Ti} = 0.1552$, and $\theta_D(\text{Sr}) = 166.5$ K, $\lambda_{Sr} = 0.146$, this leads to $T_c = 7$ K. This is indeed a rough estimate because we have not taken chemical potential into account. Nonetheless such a view of the superconductivity in SrTiO₃ assumes significance in the light of a recent observation[17] of a 400-fold increase in the conductivity of its single crystals upon exposure to sub-bandgap light (2.9 eV or higher) *at room temperature*. This phenomenon is called persistent photoconductivity because it persists for several days with negligible decay, even after the light is turned off.

Finally, we note that we have given in this chapter a direction for the solution of a long-standing puzzle — rather than a final solution.

Notes and References

1. J.F. Schooley, W.R. Hosler and M.L. Cohen, *Phys. Rev. Lett.* **12**, 474 (1964).
2. C.S. Koonce, M.L. Cohen, J.F. Schooley, W.R. Hosler and E.R. Pffiffer, *Phys. Rev.* **163**, 380 (1967).
3. D.M. Eagles, *Phys. Rev.* **178**, 668 (1969).
4. X. Lin, Z. Zhu, B. Fauqué and K. Behnia, *Phys. Rev. X* **3**, 021002 (2013).
5. G. Binnig, A. Baratoff, H.E. Hoenig and J.G. Bednorz, *Phys. Rev. Lett.* **45**, 1352 (1980).
6. A. Bussmann-Holder, A.R. Bishop and A. Simon, arxiv.org/pdf/0909.4176.
7. M. Jourdan, N. Blümer and H. Adrian, *Eur. Phys. J. B* **33**, 25 (2003).
8. X. Lin *et al.*, *Phys. Rev. Lett.* **112**, 207002 (2014).
9. D.M. Eagles, *Phys. Rev.* **186**, 456 (1969).
10. S.G. Botelho, BCS-to-BEC quantum phase transitions in high-T_C superconductors and fermionic atomic gases: A functional integral approach, *PhD thesis* (Georgia Institute of Technology, 2005).
11. P. Roth, E. Gmelin and E. Hagenbarth, in: *Phonon Scattering in Condensed Matter Vol. VII — Proceedings of the Seventh International Conference*, ed. M. Meissner and R.O. Pohl (Cornell University, Ithaca, NY, 1992).
12. J.C. Burfoot and G.W. Taylor, *Polar dielectrics and their applications* (University of California Press, 1979).
13. R.D. Parks (Ed.), *Superconductivity Vols. 1 and 2* (Marcell Dekker, NY, 1969).
14. D. van der Marel, *Physica C* **165**, 35 (1990).
15. D. van der Marel, J.L.M. van Mechelen and I.I. Mazin, *Phys. Rev. B* **84**, 205111 (2011).
16. G.P. Malik and V.S. Varma, *Int. J. Mod. Phys. B* **27**, 1350042 (2013).
17. M.C. Tarun, F.A. Selim and M.W. McClusky, *Phys. Rev. Lett.* **111**, 187403 (2013). This chapter is based on: G.P. Malik, *Int. J. Mod. Phys. B* **28**, 1450238 (2014) (14 pages).

Chapter 11

Some Exceptional Superconductors: La_2CuO_4 (LCO) and Heavy-fermion Superconductors (HFSCs)

11.1. Introduction: Features Which Make LCO and HFSCs Exceptional

11.1.1. *LCO*

It is well-known that LCO is an insulator. It becomes superconducting when suitably doped, $(La_{0.925}Sr_{0.075})_2CuO_4$ being an example which has a $T_c \approx$ 38 K. Discovered by Bednorz and Müller,[1] it occupies a unique position among all the high-temperature superconductors (HTSCs) that have been discovered so far for the following reasons:

(1) Being the first SC to transcend the BCS barrier of $T_c \approx$ 23 K, it heralded the era of high-T_c superconductivity and led to the discovery of YBCO and the Tl-, Bi- and Hg-based SCs — each of them being characterized by a T_c higher than even the liquefaction temperature of nitrogen. This of course is well known.

(2) Unlike all the HTSCs mentioned above, LCO contains predominantly only *one* species of ions that can give rise to pairing: La (strictly speaking, $La_{0.925}Sr_{0.075}$), which implies that pairing in it is governed by only one interaction parameter. This *prima facie* poses a problem because, as

was pointed out in Chapter 4, we need to employ the generalized BCS equations (GBCSEs) to deal with an SC characterized by a T_c exceeding 23 K and two gaps Δ_1 and $\Delta_2 > \Delta_1$. The application of these equations in the two-phonon exchange mechanism (TPEM) then brings into play two interaction parameters, e.g., λ_Y and λ_{Ba} for YBa_2CuO_7, and any two of λ_{Tl}, λ_{Ba}, and λ_{Ca} for $Tl_2Ba_2CaCu_2O_8$ or $Tl_2Ba_2Ca_2Cu_3O_{10}$. This is also the case for the Bi-based and iron-pnictide HTSCs, as was seen in Chapter 4.

(3) When the problem mentioned above is addressed — by appealing to the structure of LCO as will be seen below — we find that the input of its T_c *alone* yields an interaction parameter that enables one to calculate *both* its gaps. This is in contrast with all the other HTSCs for which the input of *two* parameters from the set $\{T_c, \Delta_1, \Delta_2\}$ is required to calculate the third.

This chapter addresses the need to:

(a) Bring the understanding of LCO in the framework of GBCSEs at par with that of all the other HTSCs that were dealt with in Chapter 4. This is done in the next section where for any T_c in the range of 37–40 K, it is shown in accord with experiment that $2\Delta_{20}/k_B T_c = 4.3$, and

(b) Explain *experimental* values for the gap-to-T_c ratio other than 4.3, i.e., 7.1, ≈ 8 and 9.3, attention to which was drawn by Bednorz and Müller[2] in their Nobel lecture.

We show in this chapter that an explanation of the multiple gap-values noted in (b) requires not only equations for the Δ and T_c that include chemical potential μ as a variable — equations which were employed in the study of the BCS–BEC crossover physics in Chapter 9 and the puzzle posed by $SrTiO_3$ in Chapter 10, but also a new physical idea that is inspired by the recent experimental work of Tacon *et al.*[3] This work, concerned with $YBa_2Cu_3O_{6.6}$, revealed that a very large electron–phonon coupling occurs in it in a very narrow region of phonon wavelengths. Adapting this insight for LCO, we are enabled to account for all values of its gap-to-T_c ratio.

Above features of LCO make it an exceptional HTSC.

11.1.2. *HFSCs*

HFSCs[4] belong to a fascinating family of compounds which can also be found in a variety of other states such as metallic, insulating and magnetic. They are so called because a conduction electron in them behaves as if it has an effective mass up to three orders of magnitude greater than its free mass. Superconductivity in these compounds was discovered by Steglich *et al.*[5]; since then it has been authoritatively dealt with, e.g., in Refs. 6 and 7. Some remarkable features of HFSCs are[8]:

$$E_F < k_B\theta_D, \quad k_BT_c < E_F, \quad T_c/T_F \approx T_F/\theta_D \approx 0.05, \qquad (11.1a,b,c)$$

where E_F, θ_D, T_c, and T_F denote, respectively, the Fermi energy, Debye temperature, critical temperature and Fermi temperature of the SC, and k_B is the Boltzmann constant.

Note in particular that Inequality (11.1a) is in conflict with the following basic assumption of BCS theory

$$E_F \quad \text{or} \quad \mu \gg k_B\theta_D, \qquad (11.2)$$

where μ is the chemical potential. Therefore, despite their rather *low* T_cs, HFSCs are regarded as outside the purview of the usual BCS equations. This feature — at least partly — is responsible for coinage of the term *exotic* or *unconventional* for them, making them exceptional.

HFSCs are characterized by two additional distinguishing features that have a bearing on the pairing mechanism operative in them. These are: (a) large heat capacities of conduction electrons — much larger than are found for the elemental SCs and as large as are usually associated with fixed magnetic momenta, and (b) anisotropy of their gap-structures. These features naturally led to the revival of an old idea that superconductivity may also arise without the mediating role of phonons. Thus, with He-3 in mind, triplet pairing was initially regarded as the cause of superconductivity in these SCs. However, it was subsequently shown[9–11] in three well-known papers that several experimental features of HFSCs can be explained if one assumes that magnetic fluctuations are the cause of d-wave pairing in them. By and large, this currently seems to be the popular view.

The investigation reported in this chapter is in not in a spirit of dissension with the above view. Rather, it is motivated by the question: Does the "popular" view *definitively* exclude all other mechanisms of pairing? We believe it does not because HFSCs are materials that exhibit multi-phase, multi-scale complexity. These features — together with the fact that at least one of such SCs, namely UPt_3, is already known to undergo more than one phase transition — suggest that an HFSC may well be characterized by different forms of pairing mechanisms in different parts of momentum space. This is a scenario inspired by the multi-condensate superconductivity approach advocated by Bianconi and collaborators.[12] We now recall that a case has been made[13] for the applicability of the electron–lattice interactions to a wide variety of SCs; also that, by employing such interactions, the T_cs and multiple gaps of several high-T_c SCs were accounted for by employing the generalized BCS equations in Chapter 4. Note that complexity of structure is a feature common to both high-T_c SCs and HFSCs.

Guided by above considerations, we assume here that *itinerancy* of conduction electrons and *hybridzation* of orbitals in an HFSC can cause them to be in a region of momentum space where, *effectively*, pairing is via an s-wave electron–lattice interaction. In order then to deal quantitatively with an HFSC, we need *once more* equations for its T_c and Δ_0 (the gap at $T = 0$) that are not constrained by Inequality (11.2), i.e., equations that contain μ as a variable. A feature of these equations that seems to us to have escaped serious attention is: After suitable *modification* as discussed in Sec. 11.4, they can also be employed to shed light on real life SCs such as HFSCs.

11.1.3.   *Plan for this chapter*

Our study of both LCO and HFSCs is based on the same set of *basic* equations. Since most of them have been dealt with earlier, but are scattered over chapters 3, 4, 9, and 10, we write them at one place in the next section. We begin with the equations that enable us to calculate the Debye temperatures of the ions in a sub-lattice comprised of layers designated as $A_x B_{1-x}$. These equations are followed by recalling equations for the T_c and Δ_0 of an SC in the one- and two-phonon exchange scenarios — both with and without chemical–potential. The new equations among these are the μ-incorporated equations in the two-phonon exchange scenario.

11.2. Basic Equations of this Chapter

(1) Equations for calculating the Debye temperatures of ions in a sub-lattice comprised of layers designated as $A_x B_{1-x}$ (dropping the superscript c of all variables employed in the earlier chapters and the subscript D of θ_D):

$$\theta(A_x B_{1-x}) = x\theta(A) + (1 - x)\theta(B) \tag{11.3}$$

$$\frac{\theta(A)}{\theta(B)} = \left[\frac{1 + \sqrt{m_B/(m_A + m_B)}}{1 - \sqrt{m_B/(m_A + m_B)}}\right]^{1/2}. \tag{11.4}$$

(2) One-phonon exchange mechanism (OPEM) scenario: Equations sans chemical potential μ, valid when Inequality (11.2) is satisfied:

(a) Equation for Δ_{10}:

$$\Delta_{10} = \frac{k_B \theta}{\sinh(1/\lambda)}. \tag{11.5}$$

(b) Alternative equation for Δ_{10} ($|W_{10}|$ to be identified with Δ_{10}):

$$|W_{10}| = \frac{2k_B \theta}{\exp(1/\lambda) - 1}. \tag{11.6}$$

(c) Equation for T_c:

$$1 = \lambda \int_0^{\theta/2T_c} dx \frac{\tanh(x)}{x}. \tag{11.7}$$

(3) OPEM scenario: Equations incorporating chemical potential μ, not constrained by Inequality (11.2):

(a) Equation for Δ_{10}:

$$\frac{\lambda_0}{2} I_1(\mu_0, \Delta_{10}) - \left[\frac{3}{4} I_2(\mu_0, \Delta_{10})\right]^{1/3} = 0, \tag{11.8}$$

where

$$\lambda_0 = \left[\frac{(2m/\hbar^2)^{3/2} E_F^{1/2}(0)}{4\pi^2} \right] V_0$$

$$\equiv [N(0)V]_0, \tag{11.9}$$

$$I_1(\mu_0, \Delta_{10}) = \int_{-k_B\theta}^{k_B\theta} d\xi \frac{\sqrt{\xi + \mu_0}}{\sqrt{\xi^2 + \Delta_{10}^2}}, \tag{11.10}$$

and

$$I_2(\mu_0, \Delta_{10}) = \int_{-\mu_0}^{k_B\theta} d\xi \sqrt{\xi + \mu_0} \left[1 - \frac{\xi}{\sqrt{\xi^2 + \Delta_{10}^2}} \right]$$

$$= \frac{4}{3}(\mu_0 - k_B\theta)^{3/2}$$

$$+ \int_{-k_B\theta}^{k_B\theta} d\xi \sqrt{\xi + \mu_0} \left[-\frac{\xi}{\sqrt{\xi^2 + \Delta_{10}^2}} \right]. \tag{11.11}$$

(b) Alternative equation for Δ_{10} ($|W_{10}|$ to be identified with Δ_{10}):

$$\frac{\lambda_0}{2} I_1(\mu_0, |W_{10}|) - \left[\frac{3}{4} I_2(\mu_0, |W_{10}|) \right]^{1/3} = 0, \tag{11.12}$$

where

$$I_1(\mu_0, |W_{10}|) = \int_{-k_B\theta}^{k_B\theta} d\xi \frac{\sqrt{\xi + \mu_0}}{|\xi| + |W_{10}|/2}, \tag{11.13}$$

and

$$I_2(\mu_0, |W_{10}|) = \frac{4}{3}(\mu_0 - k_B\theta)^{3/2}$$

$$+ \int_{-k_B\theta}^{k_B\theta} d\xi \sqrt{\xi + \mu_0} \left[1 - \frac{\xi}{\sqrt{\xi^2 + W_{10}^2}} \right]. \tag{11.14}$$

(c) Equation for T_c:

$$\frac{\lambda_1}{2} I_3(\mu_1, T_c) - \left[\frac{3}{4} I_4(\mu_1, T_c) \right]^{1/3} = 0, \qquad (11.15)$$

where

$$\lambda_1 = \left[\frac{(2m/\hbar^2)^{3/2} E_F^{1/2}(T_c)}{4\pi^2} \right] V_1$$

$$\equiv [N(0)V]_{T_c}, \qquad (11.16)$$

$$I_3(\mu_1, T_c) = \int_{-k_B\theta}^{k_B\theta} d\xi \frac{\sqrt{\xi + \mu_1} \, \tanh(\xi/2k_BT_c)}{\xi}, \qquad (11.17)$$

and

$$I_4(\mu_1, T_c) = \int_{-\mu_1}^{k_B\theta} d\xi \sqrt{\xi + \mu_1} \, [1 - \tanh(\xi/2k_BT_c)]. \qquad (11.18)$$

(4) TPEM scenario: Equations sans chemical potential μ, valid when Inequality (11.2) is satisfied ($|W_{20}|$ to be identified with Δ_{20}; $\Delta_{20} > \Delta_{10}$):

(a) Equation for $|W_{20}|$ (to be identified with Δ_{20}; $\Delta_{20} > \Delta_{10}$):

$$1 = \lambda_1 \ln \left[1 + \frac{2k_B\theta_1}{|W_{20}|} \right] + \lambda_2 \ln \left[1 + \frac{2k_B\theta_2}{|W_{20}|} \right]. \qquad (11.19)$$

(b) Equation for T_c:

$$1 = \lambda_1 \int_0^{\theta_1/2T_c} dx \frac{\tanh(x)}{x} + \lambda_2 \int_0^{\theta_2/2T_c} dx \frac{\tanh(x)}{x}. \qquad (11.20)$$

(5) TPEM scenario: Equations incorporating chemical potential μ, not constrained by Inequality (2):

(a) Equation for $|W_{20}|$ (to be identified with Δ_{20}; $\Delta_{20} > \Delta_{10}$):

$$\frac{\lambda_0}{2} I_1^{(2)}(\theta_1, \theta_2, \mu_0, |W_{20}|) - \left[\frac{3}{4} I_2^{(2)}(\theta_2, \mu_0, |W_{20}|) \right]^{1/3} = 0,$$

(11.21)

where

$$I_1^{(2)}(\theta_1, \theta_2, \mu_0, |W_{20}|) = \int_{-k_B\theta_1}^{k_B\theta_1} d\xi \frac{\sqrt{\xi + \mu_0}}{|\xi| + |W_{20}|/2}$$

$$+ \int_{-k_B\theta_2}^{k_B\theta_2} d\xi \frac{\sqrt{\xi + \mu_0}}{|\xi| + |W_{20}|/2},$$

(11.22)

and

$$I_2^{(2)}(\theta_2, \mu_0, |W_{20}|) = \frac{4}{3}(\mu_0 - k_B\theta_2)^{3/2}$$

$$+ \int_{-k_B\theta_2}^{k_B\theta_2} d\xi \sqrt{\xi + \mu_0} \left[1 - \frac{\xi}{\sqrt{\xi^2 + W_{20}^2}} \right].$$

(11.23)

(b) Equation for T_c:

$$\frac{\lambda_1}{2} I_3^{(2)}(\theta_1, \theta_2, \mu_1, T_c) - \left[\frac{3}{4} I_4^{(2)}(\theta_2, \mu_1, T_c) \right]^{1/3} = 0, \quad (11.24)$$

$$I_3^{(2)}(\theta_1, \theta_2, \mu_1, T_c) = \int_{-k_B\theta_1}^{k_B\theta_1} d\xi \frac{\sqrt{\xi + \mu_1} \tanh(\xi/2k_B T_c)}{\xi}$$

$$+ \int_{-k_B\theta_2}^{k_B\theta_2} d\xi \frac{\sqrt{\xi + \mu_1} \tanh(\xi/2k_B T_c)}{\xi},$$

(11.25)

and

$$I_4^{(2)}(\theta_2, \mu_1, T_c) = \int_{-\mu_1}^{k_B\theta_2} d\xi \sqrt{\xi + \mu_1} \, [1 - \tanh(\xi/2k_B T_c)].$$

(11.26)

(6) Relation between the number density of charge carriers n and Fermi energy E_F:

$$n = \frac{1}{3\pi^2}\left[\left(\frac{2m}{\hbar^2}\right)E_F(0)\right]^{3/2} \qquad (11.27)$$

Finally, we also have

$$E_F(0) = \left[\frac{\lambda_0}{2}I_1(\mu_0, \Delta_0)\right]^2, \qquad E_F(T_c) = \left[\frac{\lambda_1}{2}I_1(\mu_1, T_c)\right]^2.$$
$$(11.28a, b)$$

The new equations among the above are Eqs. (11.21) and (11.24) pertaining to TPEM scenario. These are obtained by (a) replacing the propagator V used in the OPEM scenario in Chapter 9 (Sec. 5) by the propagator $(V_1 + V_2)$, where $V_1 \neq 0$ in the energy range $-k_B\theta_1$ to $k_B\theta_1$ and $V_2 \neq 0$ in the range $-k_B\theta_2$ to $k_B\theta_2$ ($\theta_2 > \theta_1$); otherwise these parameters have vanishing values, and (b) following the procedure given after Eq. (4.9) in Chapter 4.

Undertaken below is a study of LCO and HFSCs based on the above equations.

11.3. Addressing LCO via GBCSEs

11.3.1. *Debye temperature of La ions in LCO*

Since T_c of LCO (38 K) exceeds 23 K, we must address it via GBCSEs invoking TPEM (see Chapter 4).

It is seen from Eqs. (11.3) and (11.4) that the Debye temperatures of the ions A and B depend upon their relative positions in the double pendulum (see Appendix 3A). Upon examining the structure of the unit cell of LCO we find[14] that it comprises layers of LaO, OLa, and CuO_2. This implies that if La is the lower of the two bobs of the double pendulum in the layers that comprise one of the sub-lattices, it is the upper bob in the layers of the other sub-lattice. This feature brings into play two Debye temperatures in the application of TPEM to LCO, as for any of the other HTSCs, but only one interaction parameter because it is the same species of ions in both the sub-lattices that causes pairing.

Applying Eqs. (11.3) and (11.4) to the sub-lattice of LCO comprising OLa layers in which La is the lower of the two bobs, we find that θ(OLa) $= 370\,\mathrm{K}$[14] leads to

$$\theta(\mathrm{La}) = \mathbf{104.8\,K} \quad \text{and} \quad \theta(O) = 635.2\,K \text{ (which we do not require)},$$
$$(11.29)$$

where $m_{La} = 138.91$ and $m_O = 15.999$ have been used. The results for the sub-lattice comprising LaO layers are:

$$\theta(\mathrm{La}) = \mathbf{431.1\,K} \quad \text{and} \quad \theta(O) = 308.9\,\mathrm{K}. \qquad (11.30)$$

11.3.2. Dealing with LCO with the input of its T_c via equations sans μ

Because LCO is characterized by only one interaction parameter, we put $\lambda_1 = \lambda_2 = \lambda$ in Eq. (11.29). Solution of this equation with the input of $T_c = 38\,\mathrm{K}$, $\theta_1 = 104.8\,\mathrm{K}$ from Eq. (11.29) and $\theta_2 = 431.1\,\mathrm{K}$ from Eq. (11.30) then yields

$$\lambda = 0.26818. \qquad (11.31)$$

This leads to two values for the smaller gap via Eq. (11.6) and one for the larger gap via Eq. (11.19); two values of the former are obtained because θ in Eq. (11.29) can be either 104.8 or 431.1 K. Our results then are:

$$2\Delta_{20}/k_B T_c = 4.27, \quad 2\Delta_{10}/k_B T_c = 0.27,\ 1.12. \qquad (11.32)$$

Because T_c values of LCO reported in the literature vary from 36–40 K, we have given in Table 11.1 not only the results of the above calculations for $T_c = 38\,\mathrm{K}$, but also for four other values.

11.3.3. Dealing with LCO with the input of its T_c via equations sans μ and a different a value of Debye temperature

It was mentioned above that in giving an account of the properties of LCO, Bednorz and Müller[2] in their Nobel lecture had noted that besides 4.3 for the ratio $2\Delta_0/k_B T_c$, the following values have also been reported: 7.1, ≈ 8 and 9.3. While our result for this ratio is in agreement with the value 4.3 for any T_c between 37 and 40 K, the problem

Table 11.1.   Values of the interaction parameter λ of superconducting La_2CuO_4 obtained via Eq. (11.20), and the associated gap-to-T_c ratios for different values of T_c obtained via Eqs. (11.19) and (11.6), respectively.

T_c (K)	λ	$\lvert W_{20} \rvert$ (meV)	$\dfrac{2\lvert W_{20} \rvert}{k_B T_c}$	$\lvert W_{10}(\theta_1) \rvert$ (meV)	$\dfrac{2\lvert W_{10}(\theta_1) \rvert}{k_B T_c}$	$\lvert W_{10}(\theta_2) \rvert$ (meV)	$\dfrac{2\lvert W_{10}(\theta_2) \rvert}{k_B T_c}$
36	0.26103	6.573	4.24	0.400	0.26	1.647	1.06
37	0.26461	6.784	4.26	0.422	0.26	1.737	1.09
38	0.26818	6.995	4.27	0.445	0.27	1.829	1.12
39	0.27173	7.207	4.29	0.467	0.28	1.922	1.14
40	0.27527	7.420	4.31	0.491	0.28	2.018	1.17

of explaining the larger values remains. We note in this connection that the value of θ(LCO) employed above was 370 K, whereas values both smaller[14] (360 K) and larger[15] (385 K) have also been reported in the literature. It therefore seems pertinent to investigate the extent to which the gap-to-T_c ratio changes if one were to adopt the largest of these values, i.e., 385 K. We have carried out this exercise with the following results:

$$\theta_D(\text{La, OLa sub-lattice}) = 109.1 \text{ K},$$

$$\theta_D(\text{La, LaO sub-lattice}) = 448.5 \text{ K}$$

$$T_c = 36 \text{ K} : \lambda = 0.25596, \quad \frac{2\lvert \Delta_{20} \rvert}{k_B T_c} = 4.21; \qquad (11.33)$$

$$T_c = 40 \text{ K} : \lambda = 0.26973, \quad \frac{2\lvert \Delta_{20} \rvert}{k_B T_c} = 4.28.$$

From these results we conclude that all the different observed values of $2\Delta_{20}/k_B T_c$ of LCO cannot be explained on the basis of variation in the Debye temperatures of its different samples.

This study of LCO in the framework of equations sans μ parallels earlier studies of various HTSCs in the same framework carried out in Chapter 4. Since at the end of it certain observed values of the gap-to-T_c ratio remain unaccounted for, we are led to readdress the problem in the larger framework of μ-incorporated GBCSEs for Δ and T_c. This is the task taken up in the following sections.

11.3.4. *LCO addressed via μ-incorporated GBCSEs: A consistency check of Eqs. (11.21) and (11.24)*

While equations given in Sec. 2 are perfectly general, they comprise an undetermined set. In TPEM scenario, we have only two equations, (11.21) and (11.24), containing six variables: λ_0, λ_1, μ_0, μ_1, W_{20}, and T_c. Therefore we now assume

$$\lambda_0 = \lambda_1 = \lambda, \quad \mu_0 = \mu_1 = \mu. \tag{11.34}$$

We could have made this assumption at the outset because it is in accord with a tenet of BCS theory. We chose not to do so in order to have readily available a set of equations that might be useful, should it be considered worthwhile to follow up this work with one of greater generality. All calculations in the following are carried out by assuming Eq. (11.34).

Since Eqs. (11.21) and (11.24) are the μ-incorporated versions of Eqs. (11.19) and (11.20), respectively, we need to show that when the constraint embodied in Inequality (11.2) is imposed they yield solutions in agreement with those obtained by solving the latter equations. To this end, we solve Eq. (11.24) for λ with the input of $\theta_1 = 104.8$ K, $\theta_2 = 431.1$ K [see Eqs. (11.29) and (11.30)], $T_c = 38$ K and *different values of μ*. We begin with $\mu = 300 k_B \theta_2$, which manifestly satisfies Inequality (11.2), and find that $\lambda = 0.26818$, which is the result that we had obtained earlier via Eq. (11.21) and noted in Eq. (11.31). Solution of Eq. (11.21) with $\mu = 300 k_B \theta_2$, and $\lambda = 0.26818$ then yields $|W_{20}| = 6.995$ meV, or $2|W_{20}|/k_B T_c = 4.27$, which also agrees with the result obtained earlier via Eq. (11.19) and noted in Eq. (11.32). Having thus established that Eqs. (11.21) and (11.24) satisfy the requisite consistency condition, repeating the exercise just carried out by progressively decreasing μ we find that (a) all results quoted for $\mu = 300 k_B \theta_2$ remain unchanged for values of μ up to ≈ 40 meV, and (b) for the limiting value of $\mu = k_B \theta_2 = 37.15$ meV (note that any value lower than this will cause $I_3^{(2)}(\theta_1, \theta_2, \mu_1, T_c)$ to become complex), the results are: $\lambda = 0.27507$, $|W_{20}| = 6.953$ meV, or $2|W_{20}|/k_B T_c = 4.25$.

Even after incorporating μ in the equations for T_c and Δ_0, our considerations so far have succeeded in explaining only *one* of the observed values of the gap-to-T_c ratio for LCO. Therefore there would seem to be a need for a new idea to explain the other values. It seems to us that the recent findings of Tacon *et al.*[3] provide precisely such an idea, even though they are based on experiments carried out with another HTSC ($YBa_2Cu_3O_{6.6}$). Appealing to it while solving Eqs. (11.21) and (11.24), we present in the next section a complete solution of the problem being addressed.

11.3.5. *An explanation of the different reported values of $2|W_{20}|/k_BT_c$ for LCO based on a treatment of the number equations at $T = T_c$ and $T = 0$ in the light of the experimental findings of Tacon et al.*[3]

The work reported in this section is based on an idea inspired by Tacon *et al.*'s experimental findings about the role of low-energy phonons in $YBa_2Cu_3O_{6.6}$ ($T_c = 61$ K) determined by employing high-resolution inelastic X-ray scattering. These experiments revealed features that were interpreted by the authors as signifying that (a) extremely large superconductivity-induced line-shape renormalizations are caused by phonons in a *narrow* range of momentum space and (b) the electron–phonon interaction has sufficient strength to generate various anomalies in electronic spectra, but does not contribute significantly to Cooper pairing. Having probed the electron-phonon coupling via their ingenious two-level approach of X-ray scattering, they further noted that "In terms of its amplitude, the coupling is actually by far the biggest ever observed in a superconductor, but it occurs in a *very narrow region of phonon wavelengths and at a very low energy of vibration of the atoms*." This statement induces us to review below our earlier treatment of the number equation.

We recall that the limits of the number equation — both at $T = T_c$ and $T = 0$ — were taken in the previous section to be from $-\mu$ to $k_B\theta_2$. We now call attention to the facts that (a) in dealing with an SC that has $T_c \approx 38$ K, we *need* to invoke TPEM, which (b) brings into play two Debye temperatures θ_1 and $\theta_2 > \theta_1$, and (c) if the SC were a 1-component SC characterized by θ_1, the number equation at $T = 0$ for it would comprise two terms: One corresponding to where $|W_{10}| = 0$ for electrons having energies between $-\mu$ and $-k_B\theta_1$ and the other where $|W_{10}| \neq 0$ for energies between $-k_B\theta_1$ and $+k_B\theta_1$ [see Eq. (11.11)]; similarly, if the SC were a 1-component SC characterized by θ_2, the number equation for it at $T = 0$ would comprise two terms, one corresponding to where $|W_{20}| = 0$ for energies between $-\mu$ and $-k_B\theta_2$ and the other where $|W_{10}| \neq 0$ for energies between $-k_B\theta_2$ and $+k_B\theta_2$. Note that the lower limit in both these cases is $-\mu$. In dealing with a composite SC characterized by two Debye temperatures θ_1 and $\theta_2 > \theta_1$, it was then natural to take $k_B\theta_2$ as the upper limit of the number equation, and this is the choice we had made earlier.

Guided by the insight provided by the findings of Tacon et al.[3] we now proceed to explore the consequences of narrowing down the range of phonon energies in the problem. If we require that in doing so we should keep intact for both the sub-lattices the region of momentum space for which $|W_{20}| \neq 0$, then we need to make just one change in our earlier treatment of the number equations: *drop the first term in Eq.* (11.23). We recall that this term corresponds to the region for which $|W_{20}| = 0$. We now follow up on this idea.

Our procedure above comprised solving Eq. (11.24) for λ with the input of T_c and different values of μ and then using these (λ, μ) values to solve Eq. (11.21) for $|W_{20}|$. We now adopt a variant of this procedure which comprises (a) eliminating λ from Eqs. (11.21) and (11.24) by appealing to Eq. (11.34), (b) solving the resulting equation for μ with the input of $T_c = 38$ K and $|W_{20}| = p\,k_B T_c/2$ with p as a variable, and then (c) determining λ by using either Eq. (11.21) or (11.24). This is a procedure that enables us to find quickly if sensible solutions for (λ, μ) exist for any particular value of p.

Eliminating λ from Eqs. (11.21) and (11.24) by appealing to Eq. (11.34), we have

$$\frac{\left[I_2^{(2)}(\theta_2, \mu, |W_{20}|)\right]^{1/3}}{I_1^{(2)}(\theta_1, \theta_2, \mu, |W_{20}|)} = \frac{\left[I_4^{(2)}(\theta_2, \mu, T_c)\right]^{1/3}}{I_3^{(2)}(\theta_1, \theta_2, \mu, T_c)}. \tag{11.35}$$

Attempting to solve Eq. (11.35), we find that it has no solution for $p \leq 4.2$. The solutions for μ for $4.3 \leq p \leq 9.5$ have been shown in Fig. 11.1; the corresponding values of λ obtained via Eq. (11.21) or (11.24) are given by $0.27388 \geq \lambda \geq 0.26843$. In Table 11.2, we have given the (μ, λ) solutions corresponding to the four specific values of p, i.e., 4.3, 7.1, 8, and 9.3, that were mentioned in Ref. 2. These results change insignificantly if T_c

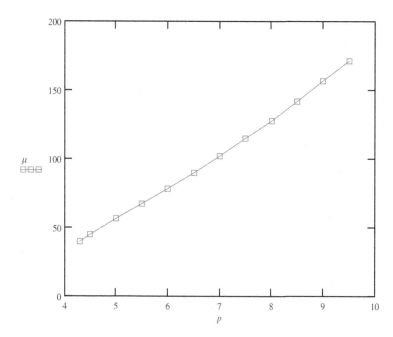

Fig. 11.1.   Plot of the chemical potential μ (meV) for different values of p obtained by solving Eq. (11.) with the input of $\theta_1 = 104.8\,\text{K}$, $\theta_2 = 431.1\,\text{K}$, $T_c = 38\,\text{K}$, and $p = 2|W_{20}|/k_B T_c$.

Table 11.2.   Values of μ obtained by solving Eq. (11.35) in the TPEM scenario with the input of $\theta_1 = 104.8\,\text{K}$, $\theta_2 = 431.1\,\text{K}$, $T_c = 38\,\text{K}$, $|W_{20}| = pk_B T_c/2$ ($p = 4.3$, 7.1, etc.) and the corresponding values of λ calculated via Eq. (11.21) or (11.24). Each pair of these (μ, λ) values is then used to calculate $|W_{10}|$ in the OPEM scenario by using Eq. (11.12), first by taking θ as 104.8 K and then 431.1 K. Adjacent to each value of $|W_{10}|$ is given in parentheses the temperature at which it is calculated via Eq. (11.15) to vanish. In the columns following these values are given the corresponding gap-to-T_c ratios with T_c taken as 38 K, i.e., *the temperature at which the larger gap vanishes.*

			OPEM scenario			
	TPEM scenario		$\theta = 104.8\,\text{K}$		$\theta = 431.1\,\text{K}$	
$p = 2|W_{20}|/k_B T_c$ ($T_c = 38\,\text{K}$)	μ (meV)	λ	$|W_{10}|$, meV (T_c, K)	$2|W_{10}|/38\,k_B$	$|W_{10}|$, meV (T_c, K)	$2|W_{10}|/38\,k_B$
4.3	39.72	0.27388	0.480 (3.1)	0.293	1.85 (11.9)	1.13
7.1	104.4	0.26886	0.449 (2.9)	0.274	1.83 (11.8)	1.12
8.0	127.8	0.26862	0.447 (2.9)	0.273	1.83 (11.8)	1.12
9.3	165.2	0.26846	0.446 (2.9)	0.273	1.83 (11.8)	1.12

is taken as 36 K or 40 K (rather than 38 K) as is seen from the following examples:

$$
\begin{array}{ccc}
 & p = 4.3 & p = 9.3 \\
T_c = 36\,\text{K}: & \mu = 40.73\,\text{meV} & \mu = 164.1\,\text{meV} \\
T_c = 40\,\text{K}: & \mu = 38.54\,\text{meV} & \mu = 166.2\,\text{meV}
\end{array}
$$

11.3.6. *Predictions of the μ-incorporated GBCSEs for LCO*

11.3.6.1. *Values of the smaller gaps*

A review of our procedure so far is as follows. By including μ in GBCSEs as applicable to LCO, appealing to an idea inspired by the work of Tacon *et al.*[3] and using as input any of the observed values of the gap-to-T_c ratio for $36 \le T_c \le 40\,\text{K}$, we have been led to solutions for μ and λ that are "sensible." Sensible because we found (a) μ to have a *low* value — in the meV range (recall that for elemental SCs, μ is in ≈ 2–$10\,\text{eV}$ range), which is in accord with the assertion made in Ref. 16 and the values reported in Ref. 17, and (b) λ to be less than 0.5, which is in accord with the Bogoliubov constraint as discussed in Chapter 4.

We now note that the (μ, λ)-values that we have been led to, enable us to calculate $|W_{10}|$ and T_c in the OPEM scenario vide Eqs. (11.18) and (11.19), respectively. These results, which constitute predictions of our approach, are also included in Table 11.2. It thus follows that, in suitably sensitized experimental set-ups, LCO should also exhibit for the gap-to-T_c ratio (with T_c taken as 38 K) the following values: 0.27 – 0.29 ($\approx 3\,\text{K}$) and 1.12 ($\approx 12\,\text{K}$), where the numbers in the parentheses denote the temperatures at which these smaller gaps are predicted to vanish.

11.3.6.2. *Carrier concentration*

We can *estimate* carrier concentration n via (11.27) by replacing E_F in it by μ. However, before we can use this equation we need to find the band effective mass of electrons in LCO. This can be done by using the following expressions for N(0) as in $\lambda = [N(0)V]$ [see Eqs. (11.9) or (11.16)]:

$$
N(0) = \frac{1}{4\pi^2} \left(\frac{2 s m_e}{\hbar^2} \right)^{3/2} \mu^{1/2}, \tag{11.36}
$$

where we have put the band effective mass as s times the free electron mass m_e, and[18]

$$N(0) = \frac{3\gamma}{2\pi^2 k_B^2 v} \equiv \rho, \tag{11.37}$$

where γ is the experimentally obtained electronic specific heat constant (also known as the Sommerfeld constant) and v the gm-at volume.

From Eqs. (11.36) and (11.37) we obtain

$$s = \frac{\hbar^2}{2m_e} \left(\frac{4\pi^2 \rho}{\mu^{1/2}} \right)^{2/3}. \tag{11.38}$$

We now use[14]: $\gamma = 4.5 \, \text{mJ/mol} \, K^2$, $v(\text{La}) = 22.60 \, \text{cm}^3$ and the values of μ corresponding to $p = 4.3$ and 9.3 to obtain

$$s(\mu = 39.72 \, \text{meV}) = 11.2, \quad s(\mu = 165.2 \, \text{meV}) = 6.96 \tag{11.39}$$

Using these values in Eq. (11.36) we obtain

$$\text{For } \mu = 39.72 \, \text{meV}, n = 1.35 \times 10^{21} cm^{-3}$$
$$\text{For } \mu = 165.2 \, \text{meV}, n = 5.60 \times 10^{21} cm^{-3}. \tag{11.40}$$

Since the value of carrier concentration noted in Ref. 2 is "of the order of $10^{21}/\text{cm}^3$," these results support the approach followed in this study.

11.4. Addressing HFSCs via BCS Equations

In the overview of HFSCs given in Sec. 11.1.2, it was pointed out that superconductivity in these is generally believed to be driven by d-wave pairing caused by magnetic fluctuations. This is a view made popular by three well known papers.[9–11] Because of reasons stated in the same section, we now address HFSCs via the more familiar s-wave pairing. Motivation for undertaking such an exercise is provided by the premise that the s- and the d-wave forms of pairing may well be considered as supplementing each other, rather than being exclusive. A signature of this being so would be the exhibition of more than one phase transition by HFSCs — a feature that at least one HFSC, namely UPt$_3$, is known to exhibit. In the following, we deal with the HFSC that has the highest T_c in its family, i.e., CeCoIn$_5$.

11.4.1. *Pairing via Ce ions in CeCoIn₅: Debye temperature of Ce ions*

Characteristic features of $CeCoIn_5$ of immediate concern to us are:

Structure: layers of $CeIn_3$ and $CoIn_2$;

$$T_c = 2.3 \text{ K};^{19} \quad \text{and} \quad \theta(CeCoIn_5) = 161 \text{ K}.^{20} \tag{11.41}$$

Because T_c of $CeCoIn_5$ is less than 23 K, we can deal with it via *modified* BCS equations, as will shortly be discussed. It follows from the structure of $CeCoIn_5$ that pairing in it via the OPEM scenario — which is the mechanism underlying BCS equations — can be caused by either of Ce and Co ions. We first consider Ce ions.

Assuming that Ce is the *lower* of the two bobs in the double pendulum, we now use Eqs. (11.3) and (11.4) with $x = 0.75$, $\theta(In_{0.75}Ce_{0.25}) = 161$ K, $m(In) = 114.82$ and $m(Ce) = 140.12$ to obtain

$$\theta(In) = 190.2 \text{ K (which we do not require), and } \theta(Ce) = \textbf{73.3 K}. \tag{11.42}$$

The assumption that Ce is *upper* of the two bobs in the double pendulum leads to

$$\theta(Ce) = \textbf{276.3 K}, \quad \text{and} \quad \theta(In) = 122.6 \text{ K (which we do not require)}. \tag{11.43}$$

In the following we shall perform all calculations with both the values of $\theta(Ce)$ noted in Eqs. (11.42) and (11.43).

11.4.2. *Modification of BCS equations to deal with HFSCs*

An important feature of the μ-incorporated Eqs. (11.8) and (11.15) is that expressions for $I_1(\mu, \Delta_{10})$, $I_2(\mu, \Delta_{10})$ and $I_3(\mu_1, T_c)$ in these, given by Eqs. (11.10), (11.11) and (11.17), respectively, lead to complex values when $\mu < k\theta$. Real solutions can however still be obtained by replacing the lower limit $(-k\theta)$ in each of these equations by $-\mu$. It is in this sense that the word *modification* was used above. While the issue of such solutions will be further discussed below, we shall as before refer to any equation with curtailed limits as a *truncated* equation.

11.4.3. *A consistency check of Eqs. (11.8) and (11.15)*

In dealing with $CeCoIn_5$, keeping Inequality (11.1a) in view, we need to solve Eqs. (11.8) and (11.15) for $\mu < k_B\theta(Ce)$. Before we do so, as a consistency check, we solve these equations for a value of μ that satisfies Inequality (11.2).

Choosing $\mu = 100\,k_B\theta$ (Ce) — which manifestly satisfies Inequality (11.2) — and using as input $T_c = 2.3$ K and $\theta(Ce) = 73.3$ K into Eq. (11.15), we obtain $\lambda = 0.2788$. Employing these values of μ, $\theta(Ce)$ and λ as input, solution of Eq. (11.8) then gives $\Delta_0 = 3.50 \times 10^{-4}$ eV. Upon solving the usual BCS Eqs. (11.7) and (11.5) — which do not require the input of μ — these are also the values for λ and Δ_0 that we obtain for the same values of T_c and $\theta(Ce)$. Changing $\theta(Ce)$ to 276.3 K and solving Eq. (11.15) changes the value of λ to 0.2035; solution of Eq. (11.18) then leads to the same value of Δ_0 that was obtained earlier. These results, i.e., the values of λ and Δ_0, are also obtained by solving the usual BCS equations with the same inputs ($T_c = 2.3$ K, $\theta(Ce) = 276.3$ K).

All results of the preceding paragraph remain valid for values of μ greater than $100\,k_B\theta(Ce)$. It is hence seen that, when Inequality (11.2) is imposed, the μ-incorporated Eqs. (11.15) and (11.8) yield solutions in accord with solutions of the usual BCS equations — as was to be expected.

11.4.4. *Solutions of Eqs. (11.15) and (11.8) with the input of θ(Ce) = 73.3 K, T_c = 2.3 K, and different values of μ*

Combining Inequalities (11.1a) and (11.1b), we obtain

$$k_B\theta(Ce) > \mu > k_BT_c(6.317 \times 10^{-3} > \mu > 1.982 \times 10^{-4}\,eV). \quad (11.44)$$

We now solve Eqs. (11.15) and (11.8) for values of μ in a range that marginally exceeds the one specified by Inequalities (11.44), i.e., $1.05\,k_B\theta(Ce) \geq \mu \geq 0.95\,k_BT_c$ ($6.632 \times 10^{-3} \geq \mu \geq 1.883 \times 10^{-4}$ eV). This is done in order to bring out the change that occurs when μ is lowered across $k_B\theta(Ce)$. For *decreasing* values of μ, we proceed as follows:

(a) Solve Eq. (11.15) for λ beginning with $\mu = 1.05\,k_B\theta(Ce)$.
(b) Solve Eq. (11.8) for Δ_0 with the input of (μ, λ) values from (a).
(c) Repeat the above steps for $\mu = k_B\theta(Ce)$.

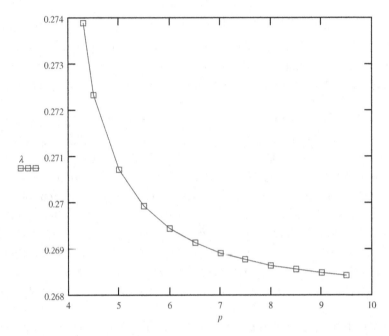

Fig. 11.2.　Plot of the coupling strength λ for different values of p obtained by solving (a) Eq. (11.35) for μ with the input of $\theta_1 = 104.8\,\text{K}$, $\theta_2 = 431.1\,\text{K}$, $T_c = 38\,\text{K}$, $p = 2|W_{20}|/k_B T_c$, and then (b) using these values in Eq. (11.21) or (11.24) to solve for λ.

(d) Repeat steps (a) and (b) for values of $\mu < k_B\theta(\text{Ce})$. Upon attempting to do so, Eq. (11.10), Eqs. (11.11) and (11.17) are found to lead to complex-valued expressions because of the factor $\sqrt{\xi + \mu}$ in each one of them. Hence, as was pointed out earlier, real solutions are now obtained via *truncated* equations.

(e) Repeat steps (a) and (b) for 13 values of μ in the range 0.9 $k_B\theta(\text{Ce})$ $\geq \mu \geq 0.95 k_B T_c$ by employing truncated equations.

The results of these calculations have been shown in the plots of λ and Δ_0 against μ in Figs. 11.3 and 11.4, respectively. For convenience of the reader, numerical values corresponding to four special points in these figures have been given in Table 11.3. These points are (i) $\mu = k_B\theta(\text{Ce})$ (6.317 meV), i.e., the end-point on the LHS of Inequality Eq. (11.44), (ii) the point corresponding to apex of the μ versus λ plot, i.e., $\mu \approx 0.3 k_B\theta(\text{Ce})$ (1.895 meV), (iii) the point where $T_c/T_F \approx T_F/\theta(\text{Ce})$ (see Eq. 11.1c), i.e., $\mu = 1.119\,\text{meV}$, and (iv) $\mu = k_B T_c$ (1.982 × 10⁻⁴ eV), i.e., the end-point

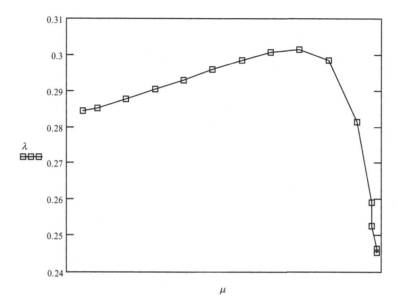

Fig. 11.3.   For pairing via the sub lattice containing Ce ions, plot of λ versus μ obtained by solving Eq. (11.15) with the input of $\theta_D = 73.3$ K and $T_c = 2.3$ K for *decreasing* values of μ in the range $1.05 \, k_B\theta_D \geq \mu \geq 0.95 k_B T_c$.

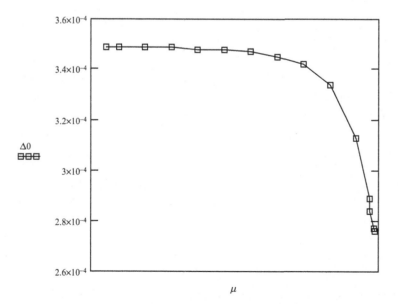

Fig. 11.4.   For pairing via the sub lattice containing Ce ions, plot of Δ_0 versus μ obtained by solving Eq. (11.12) corresponding to values of θ_D, μ, and λ in Fig. 11.3.

Table 11.3. For pairing via Ce ions, values of λ obtained by solving *truncated* Eq. (11.15) for four select values of μ with the input of $T_c = 2.3$ K and two values of θ_D, i.e., **73.3 K**, and **276.3 K**. Values of (μ, λ) in each row are then used to calculate Δ_0, $E_F(0)$, and $E_F(T_c)$ via Eqs. (11.12), (11.28a), and (11.28b), respectively. The values of s are obtained by using Eq. (11.46).

						Remark (Results correspond to the value of μ given by)
μ (meV)	λ	$\Delta_0(10^{-4}$ eV)	$E_F(0)$ (meV)	$E_F(T_c)$ (meV)	s	
$\theta_D(\text{Ce}) = 73.3$ K						
$k_B\theta_D$ (6.317)	0.2853	3.49	6.332	6.322	94	LHS of (21)
$0.3\,k_B\theta_D$ (1.895)	0.3016	3.42	1.943	1.912	140	Apex of the plot in Fig. 1
$k_B\sqrt{T_c}\theta_D$ (1.119)	0.2964	3.31	1.189	1.149	166	$T_c/T_F \approx T_F/\theta_D$; See Eq. (11.1c)
$k_B T_c$ (0.198)	0.2466	2.77	0.350	0.324	253	RHS of Eq. (11.21)
$\theta_D(\text{Ce}) = 276.3$ K						
$k_B\theta_D$ (23.81)	0.2069	3.49	23.82	23.81	60	LHS of (21)
$0.3\,k_B\theta_D$ (7.143)	0.2149	3.48	7.162	7.147	90	Apex of the λ versus μ Plot
$k_B\sqrt{T_c}\theta_D$ (2.172)	0.1980	3.37	2.222	2.187	134	$T_c/T_F \approx T_F/\theta_D$; See Fig. (11.1c)
$k_B T_c$ (0.198)	0.1213	2.52	0.342	0.324	253	RHS of Eq. (11.19)

on the RHS of Inequality (11.44). Corresponding to each pair of (λ, Δ_0) values in Table 11.3 are also given values of $E_F(0)$ and $E_F(T_c)$ which are obtained via Eqs. (11.28a) and (11.28b), respectively.

11.4.5. *Effective mass of conduction electrons*

Given the value of $E_F(T_c)$, the effective mass of conduction electrons can be calculated by appealing to two alternative expressions for N(0). One of these, as seen from Eq. (11.16), is:

$$N(0) = \frac{1}{4\pi^2}\left(\frac{2s\,m_e}{\hbar^2}\right)^{3/2} E_F^{1/2}(T_c), \qquad (11.45)$$

where we have put the band effective mass as s times the free electron mass m_e.

Table 11.4. For pairing via Co ions, values of λ obtained by solving *truncated* Eq. (11.15) for four select values of μ with the input of $T_c = 2.3$ K and two values of θ_D, i.e., **98.7 K**. and **294.7 K**. Values of (μ, λ) in each row are then used to calculate Δ_0, $E_F(0)$, and $E_F(T_c)$ via Eqs. (11.12), (11.28a), and (11.28b), respectively. The values of s in the last column are obtained by using Eq. (11.46).

			$\theta_D(Co) = 98.7$ K		
μ(meV)	λ	Δ_0 (10^{-4} eV)	$E_F(0)$ (meV)	$E_F(T_c)$ (meV)	s
$k_B\theta_D$ (8.505)	0.2629	3.49	8.518	8.509	181
$0.3k_B\theta_D$ (2.552)	0.2764	3.44	2.591	2.564	270
$k_B\sqrt{T_c\theta_D}$ (1.298)	0.2683	3.31	1.364	1.324	337
k_BT_c (0.198)	0.2096	2.71	0.348	0.324	538
			$\theta_D(Co) = 294.1$ K		
$k_B\theta_D$ (25.34)	0.2043	3.49	25.35	25.34	126
$0.3\,k_B\theta_D$ (7.603)	0.2120	3.48	7.621	7.603	188
$k_B\sqrt{T_c\theta_D}$ (2.241)	0.1946	3.37	2.290	2.256	282
k_BT_c (0.198)	0.1174	2.52	0.342	0.324	538

Another expression for N(0) was given above in Eq. (11.37). From Eqs. (11.45) and (11.37) we obtain

$$s = \frac{(\hbar c)^2}{2m_e c^2}\left[\frac{4\pi^2\rho}{E_F^{1/2}(T_c)}\right]^{2/3}. \tag{11.46}$$

The values of γ is:[19]

$$\gamma = 0.04 \text{ J mol}^{-1}\text{K}^{-2}. \tag{11.47}$$

Using Eq. (11.47) and $v(Ce) = 20.69$ cm^3 mol^{-1}, ρ can be calculated via Eq. (11.37). On doing so, the values of s obtained via Eq. (11.46) for the four special points in the λ vs. μ plot in Fig. 11.1 that were mentioned above have also been included in Table 11.1.

11.4.6. Solutions of Eqs. (11.15) and (11.8) with the input of θ (Ce) = 276.3 K, T_c = 2.3 K, and different values of μ

Combining Inequalities Eq. (11.1a) and (11.1b), we now have

$$k_B\theta(Ce) > \mu > k_BT_c(23.8 \times 10^{-3} > \mu > 1.982 \times 10^{-4}eV). \tag{11.48}$$

As we did for $\theta(\text{Ce}) = 73.3\,\text{K}$, we now solve Eqs. (11.15) and (11.8) for 15 values of μ in the range $1.05\,k_B\theta(\text{Ce}) \geq \mu \geq 0.95\,k_B T_c$ ($25.0 \times 10^{\times 3} \geq \mu \geq 1.883 \times 10^{-4}$ eV), which marginally exceeds the range given by Inequalities Eq. (11.48). Since the plots of λ and Δ_0 against μ are found to be similar to those shown in Figs. 11.1 and 11.2, respectively, they are not shown here. However, we have included in the lower part of Table 11.1 all the results akin to those that were given in the upper part of the table for $\theta(\text{Ce}) = 73.3\,\text{K}$.

11.4.7.   *Pairing via with the sub-lattice containing Co ions*

We have so far considered the effects of pair-formation via the Ce ions. On repeating the above exercise with Co ions, we find that $\theta(\text{Co}) = \textbf{98.7 K}$ in In_2Co layers, i.e., when Co is lower of the two bobs in the double pendulum; otherwise $\theta(\text{Co}) = \textbf{294.1 K}$ (in CoIn_2 layers). On now solving Eqs. (11.4) and (11.3), the plots of λ and Δ_0 against μ are found to be similar to those obtained earlier and hence are not included here. Given in Table 11.2 are all the other results akin to those that were given in Table 11.2 for pairing via the Ce ions. We note that the value of $v(\text{Co})$ is $6.67\,\text{cm}^3$, which differs considerably from that of $v(\text{Ce})$.

11.4.8.   *Density of charge carriers*

Density of charge carriers n can be calculated by using Eq. (11.27). The results following from the s and $E_F(T_c)$ values given in the two Tables are as follows:

For pairing via the Ce ions:

$$7.84 \times 10^{21}[E_F(T_c) = 23.81\,\text{meV}, s = 60.4]$$
$$\geq n(\text{cm}^{-3}) \geq 1.07 \times 10^{20}[E_F(T_c) = 0.324\,\text{meV}, s = 253]$$

For pairing via the Co ions:

$$2.59 \times 10^{22}[E_F(T_c) = 25.34\,\text{meV}, s = 126]$$
$$\geq n(\text{cm}^{-3}) \geq 3.31 \times 10^{20}[(E_F(T_c) = 0.324\,\text{meV}, s = 539]$$

11.5. Remarks

1. While CeCoIn$_5$ (T_c = 2.3 K) was addressed in this chapter via the μ-incorporated *truncated* BCS equations which are based on OPEM scenario, LCO was dealt with via μ-incorporated *truncated* GBCSEs in the TPEM scenario because its T_c exceeds 23 K. Further, while truncation in each case reduces the range of energies for electrons to be paired via the electron–lattice interactions, the precise amounts by which these energies were curtailed were different in the two cases — while for LCO it sufficed to drop the first term in Eq. (11.23), for CeCoIn$_5$ curtilment of these energies was more severe.

2. We recall that we came across the need to *truncate* BCS equations for the first time in Chapter 9, where the crossover problem was addressed without appealing to SLT for the hypothetical case of the *element* Sn. *Truncation* was also resorted to while dealing with the puzzle posed by SrTiO$_3$ in Chapter 10, where it was pointed out that — apart from the straightforward need to avoid comlex-valued solutions — it may also be justified on physical grounds by imagining electrons to be, as it were, in a "rain-shadow region" of reduced pairing. Having invoked this concept in three successive chapters, there would seem to be a need for further elaboration. Also because our treatment of LCO based on TPEM, and of CeCoIn$_5$ on OPEM, is likely to seem rather naive in view of the known facts that HTSCs and HFSCs are character-ized by multi-phase and multi-scale complexity and that their Fermi surfaces have highly complex structures comprising many sheets that span different bands. We are therefore impelled to draw attention to the following:

(a) As can be seen in Ref. 21, Fermi surfaces of elemental SCs too have rather complex structures: **none** of them has the idealized spherical Fermi surface assumed in BCS theory. And yet, with the exception of a few, e.g., Pb and Hg, the mean-field approximation (MFA) employed in the theory works for most of them.

(b) Itinerancy of electrons in a multi-band SC is a much-invoked concept since at least the time of Suhl *et al.*'s paper.[22] We give below an analogy to bring out how this concept justifies *dropping of the first term* in the number Eq. (11.23) and curtailment of the lower limits of equations dealing with CeCoIn$_5$.

(c) Consider a convoy on a road passing through a range of mountains. As the road twists and turns through a series of valleys and mountains, the amount of sunlight it receives will vary from a maximum at the highest point of the range to a minimum at a place determined by the topography of the entire range. Streams of *itinerant* conduction electrons in a solid on a 3-d Fermi surface are akin to such a convoy: there will be places where they can exchange phonons with the A- or the B-ions separately, and a place or places where they can exchange phonons with both of them simultaneously. For the last of these cases, one can then envisage for LCO a situation where phonon energies span not the usual range from $-\mu$ to $k_B\theta_2$, but a depleted range from $-k_B\theta_2$ to $k_B\theta_2$. The *mean* value of μ for the electrons under consideration is equivalent to the word "place" used for the convoy. Curtailment of limits of the equations dealing with CeCoIn$_5$ can be similarly justified.

(d) It is evident that the values of μ corresponding to different values of Δ_2 and T_c that we have determined for LCO reflect the structure of the Fermi surface of the SC. Implicit to our approach is the concept of *locally spherical Fermi surfaces* (LSFSs), which takes complexity of the HTSC into account in a rather simple (if not the simplest possible) manner. Such simplicity of approach for the kind of system we are dealing with is a goal every model strives to achieve. We note that LSFSs come into play because of itinerancy of electrons and the adoption of MFA.

(e) It seems interesting to point out that the concept of LSFSs is similar to the concept of *locally inertial coordinate frames* employed in the general theory of relativity.[23]

(f) We should like to note that the concept of LSFSs also underlies the explanation of the multiple gaps of various HTSCs as was noted in Chapter 4: they correspond to different values of E_F. This is not immediately evident because the values of E_F are not required for their calculation, just as they are not required for the elemental SCs because of the assumption $E_F \gg k_B\theta_D$.

11.5.1. *LCO: discussion*

1. Strictly speaking, Fermi energy is a term applicable to non-interacting systems. In the study above concerned with LCO, however, it was used interchangeably with the chemical potential, as is usually done in the literature on superconductivity.

2. In connection with the remark by Tacon *et al.*[3] that "the electron–phonon interaction has sufficient strength to generate various anomalies in electronic spectra, but does not contribute significantly to Cooper pairing", we should like to note that this statement is presumably based keeping OPEM in mind, which cannot account for the observed T_c of LCO.

3. The values of $\gamma = 4.5$ mJ/mol K^2 and v = 22.60 cm^3 used on page 179 warrant a comment. The former of these pertains to the electron concentration in the *composite* LCO, pairing of which causes superconductivity. The latter value, on the other hand, is for La in its free state. We note that, based on the crystallographic data, v for LCO has approximately the value 8.14 cm^3 (in the orthorhombic state; LCO also exists in tetragonal state for which v has nearly the same value). Use of this value would imply that, at the micro level, the Cu and O ions play a direct role in the scatterings responsible for pairing, whereas this role must be restricted to the positively charged ions. While indeed the use of the γ value for free La is an approximation, it seems to be justified by the results it has led to. This comment also applies to our treatment of CeCoIn$_5$: employing the v value for this *compound* calculated by using crystallographic data would imply that there is no difference (other than the difference in their Debye temperatures) between scatterings due to the Ce and the Co ions that cause pairing.

11.5.2. *HFSCs: discussion*

1. Based on experiments performed by employing the Bogoliubov quasiparticle interference (QPI) technique, it has recently been reported[24] that CeCoIn$_5$ is characterized by a multitude of gap-values, the greatest among these being $(5.5 \pm 0.05) \times 10^{-4}$ eV. This value has been specifically attributed to the region of $d_{x^2-y^2}$ pairing. The investigation reported here too has led to a multitude of Δ_0 values — in the $(2.76 - 3.49) \times 10^{-4}$ eV range for pairing via the Ce ions with

$\theta_D = 73.3$ K — corresponding to a single T_c value. The fact that our maximum value is lower than the maximum value reported in[24] is not surprising in view of the mean-field approximation on which it is based.

2. That there *must* be an attribute of CeCoIn$_5$ which causes it to have the above features (single T_c, multiple gaps) is obvious. By and large, μ is believed to be that attribute because Δ_0 is supposed to follow its fluctuations. The new feature of our approach is the suggestion that HFSCs be addressed by incorporating μ into the BCS equations for T_c and Δ_0. Implementation of this suggestion led us to consideration of four equations: the μ-incorporated BCS equation for T_c and the corresponding number equation, besides the μ-incorporated BCS equation for Δ_0 and the number equation at $T = 0$. While the latter two equations have provided the starting point of *all* papers that have addressed the problem of the BCS–BEC crossover physics, what seems to us to have escaped serious attention is that, upon *modification*, they can also be used to account for the properties of HFSCs.

3. For *low* values of μ, the difference between these and the corresponding values of $E_F(0)$ and $E_F(T_c)$ in Tables 11.3 and 11.4 is rather pronounced. This warrants a comment because not only are μ and E_F generally used interchangeably in the superconductivity literature, but also because no distinction is made between the values of E_F at $T = 0$ and $T = T_c$. While both E_F and μ signify the amount of energy required to add one particle to the existing many-particle system, strictly speaking, the former is a concept applicable to a non-interacting system whereas the latter applies to an interacting system. Since temperature lifts the degeneracy of a non-interacting system of fermions at $T = 0$, it follows that $E_F(0)$ should be greater than $E_F(T_c)$, as is also seen from the two Tables. Further, because interactions cause a reduction in the number of free electrons because of pairing, it follows that μ should be less than both $E_F(0)$ and $E_F(T_c)$, as is also reflected in the two Tables. While we have so far talked about *low* values of μ, we note that the inequalities $E_F(0) > E_F(T_c) > \mu$ hold for *any* value of μ, as can be seen even from the top row in Table 11.1. We also note that for $\mu \gg k_B\theta_D$, the differences between them show up in the fifth or sixth places after the decimal.

4. We now draw attention to Figs. 11.3 and 11.4. The first of these has a dome-like structure, which implies that one can have the same value of λ for two different values of μ. The import of the second figure is that the *same* value of λ — when combined with different values of μ — leads to different values of Δ_0. We note that the parabolic feature of Fig. 11.4 brings out that the variation in Δ_0 for *decreasing* values of μ (for $T = T_c$) is similar to its variation for *increasing* values of T (when $\mu \gg k_B \theta_D$).

5. We should finally like to recall that incorporation of an external magnetic field via the Landau quantization scheme into the BCS equations enabled us in Chapter 7 to shed light on the de Haas–van Alphen oscillations. A further extension of this work incorporating chemical potential may well lead to additional insights about HFSCs.

6. To conclude, we have shown in this chapter that even the so-called exotic or unconventional SCs can be brought into the purview of an extended form of BCS theory which is based significantly on the concept of MDTs due to Born and Karmann as discussed in Chapter 4, Suhl *et al.*[22], and Bianconi *et al.*[12] The application of the modified theory to CeCoIn$_5$ has shed light on not only its T_c(s) and multiple gaps, but also on the mass values of its *heavy* electrons and their number densities.

Notes and References

1. J.G. Bednorz and K.A. Müller, *Z. Phys. B* **64**, 189 (1986).
2. J.G. Bednorz and K.A. Müller, *The New Approach to High-T_c Superconductivity*, Nobel Lecture, 424 (1987).
3. M. Le Tacon *et al.*, *Nature Phys.* **10**, 52 (2014).
4. A. C. Hewson, *The Kondo Problem to Heavy Fermions, Cambridge Studies in Magnetism (No. 2)* (Cambridge University Press, Cambridge, 1993).
5. F. Steglich *et al.*, *Phys. Rev. Lett.* **43**, 1892 (1976).
6. P.S. Riseborough, G.M. Schmiedeshoff, and J.L. Smith, *Heavy Fermion Superconductivity, Springer Series in Solid State Sciences* (Springer Verlag, Berlin, 2008).
7. P. Coleman, *Heavy Fermions: electrons at the edge of magnetism*, in *Handbook of Magnetism and Advanced Magnetic Materials,* Vol. 1, Eds. H. Kronmuller and S. Parkin (John Wiley, Chichester, 2007).
8. J.D. Thompson, Talk given at the Conference, BCS Theory @ 50, University of Illinois at Urbana-Champaign, Oct 10–13, 2007.
9. K. Miyake, S.S. Rink and C.M. Varma, *Phys. Rev. B* **34**, 6554 (1986).
10. M.T.B. Monod, C. Bourbonnias and V. Emery, *Phys. Rev. B* **34**, 7716 (1986).
11. D.J. Scalapino, E. Loh and J.E. Hirsch, *Phys. Rev. B* **34**, 8190 (1986).
12. A. Bianconi, *Nat. Phys.* **9**, 536 (2013).

13. V.Z. Kresin and S.A. Wolf, *Rev. Mod. Phys.* **81**, 481 (2009).
14. C.P. Poole Jr., H.A. Farach, R.J. Creswick and R. Prozov, *Handbook of Superconductivity*, 2nd edn. (Academic Press, Amsterdam, 2007)
15. K. Berggold *et al.*, *Phys. Rev. B* **73**, 104430 (2006).
16. A.S. Alexandrov, *Nonadiabatic superconductivity in MgB_2 and Cuprates* (2001). http://arxiv.org/abs/cond-mat/0104413
17. T. Jarlborg and A. Bianconi, *Phys. Rev. B* **87**, 054514 (2013).
18. D. Pines, *Phys. Rev.* **109**, 280 (1958).
19. C. Petrovic *et al.*, *J. Phys.: Condens. Matter* **13**, L337 (2001).
20. J.S. Kim *et al.*, *Phy. Rev. B* **64**, 134524 (2001).
21. A.P. Cracknell and K.C. Kong, *The Fermi Surface* (Clarendon Press, Oxford, 1973).
22. H. Suhl, B.T. Matthias and L.R. Walker, *Phys. Rev. Lett.* **3**, 52 (1959).
23. S. Weinberg, *Gravitation and Cosmology: Principles and Applications of the General Theory of Relativity* (John Wiley & Sons, NY, 1972).
24. M.P. Allan, *Nat. Phys.* **9**, 468 (2013).
    This chapter is based on the following papers:
    G.P. Malik and V.S. Varma, *WJCMP* **5**, 148 (2015), where LCO has been dealt with; and
    G.P. Malik, *J. Mod. Phys.* **6**, 1233 (2015), where HFSCs have been dealt with.

Appendix 11A: A review of the procedure for calculating Debye temperatures of A and B ions, given the Debye temperature of the sub-lattice comprising layers of $A_x B_{1-x}$

$CeCoIn_5$ consists of sub-lattices that contain layers of $CeIn_3$ and $CoIn_2$. Any layer in each of these sub-lattices was assumed to have the Debye temperature of $CeCoIn_5$. Assuming further that Ce and In ions in a $CeIn_3$ layer simulate the weakly coupled oscillations of a double pendulum, their Debye temperatures were calculated by employing the following equations:

$$\theta(A_x B_{1-x}) = x\theta(A) + (1 - x)\theta(B) \qquad (A.11.1)$$

$$\frac{\theta(A)}{\theta(B)} = \left[\frac{1 + \sqrt{m_B/(m_A + m_B)}}{1 - \sqrt{m_B/(m_A + m_B)}} \right]^{1/2}. \qquad (A.11.2)$$

Since we do not know whether Ce is the lower or the upper bob in the double pendulum, we allowed for both these possibilities. With Ce regarded as the lower bob, we have A = In, B = Ce, $x = 3/4$; the input of $\theta(A_x B_{1-x}) = 370$ K and the masses of Ce and In ions then led to $\theta(Ce) = $ **73.3** K. The other choice (Ce regarded as the lower bob) led to $\theta(Ce) = $ **276.3** K. Similarly, for pairing via the Co ions, we were led to $\theta(Co) = $ **98.7** K and **294.1** K.

Characterization of an anisotropic non-elemental SC by more than one Debye temperature is an idea that was first proposed by Born and Karmann in a different context, as was discussed in Chapter 4. While we have implemented this idea via the above equations for all the CSs dealt with in this monograph, we should like to note that, as was also pointed out in Ref. 25c, our procedure is meant to be suggestive. There may well be other ways in which anisotropy of these SCs can be taken into account. As an example, one could replace Eq (A.11.2) by

$$\frac{\theta(A)}{\theta(B)} = \sqrt{\frac{m_b}{m_a}}, \qquad (A.11.3)$$

which leads to a different set of values for $\theta(Ce)$ and $\theta(Co)$ than found earlier. Hence the question: how is one to distinguish between different θ-values of the same species of ions? In dealing with $CeCoIn_5$, we found that its T_c and Δ_0 can be explained by employing such disparate θ values as 73.3 and 276.3 K (in the OPEM scenario for pairing via the Ce ions). This is so because any change in θ causes a corresponding change in the interaction parameter λ, and different sets of $\{\theta, \lambda\}$ values can account for the same values of T_c and Δ_0. In this sense the properties of the SC seem not to be sensitively dependent on the choice of θs. However, different values of θ *finally* led to different values for both the effective mass of conduction electrons and the number density of charge carriers. Hence we can rule out one of these θ values by be monitoring, e.g., the latter of these parameters via the Hall effect.

In dealing with an SC characterized by $\{T_c, \Delta_1, \Delta_2\}$ in the TPEM scenario, the choice of different sets of $\{\theta_1, \theta_2\}$-values will lead, with the input of $\{T_c, \Delta_2\}$, to different sets of values of $\{\lambda_1, \lambda_2\}$. Calculation of Δ_1 via (λ_1, θ_1) and (λ_2, θ_2) can then shed light on which of the $\{\theta_1, \theta_2\}$ sets is operative in the SC.

Experimental verification of the idea that constituents of a CS have different Debye temperatures has been provided by Kwei and Lawson, *Physica C* **175**, 135 (1991). Specifically, based on neutron powder diffraction experiments, they have reported that LCO is characterized by the following anisotropic values: $\theta_{xx} = 238, \theta_{yy} = 267, \theta_{zz} = 349$ (K). Significantly, they have also suggested that these values are temperature-dependent. It then follows that, as discussed in this Chapter and earlier in Chapter 10, the interaction parameters in the theory should also be regarded as T-dependent. Confirmation of this idea too has recently been provided by the experimental work of Mansart *et al.* (*Phys. Rev. B* **88**, 054507, 2013). This was briefly discussed in the final Remark in Chapter 5.

Chapter 12

Solar Emission Lines and Quarkonium Mass Spectra

12.1. Introduction

The topics covered in this chapter might be found surprising because *prima facie* there is nothing common between any of them and superconductivity. However, we draw attention to the fact that each of these topics is concerned with *a bound state problem in a medium at finite temperature*: solar emission lines with the bound states of all kinds of ions, spectra of mesons with the bound states of appropriate quarks, and superconductivity with the bound states of electrons. Further, for reasons given below, if one should attempt to account for the hosts of solar emission lines that remain unidentified to date on the basis of transitions in a single system comprised of an electron and a proton at different temperatures, then one is led back to the BSE + Matsubara approach followed for superconductivity, albeit with a different kernel. The observed masses of various mesons and their de-confinement temperatures can similarly be addressed by an appropriate choice of the kernel via the same approach. It should hence be seen that BSE + Matsubara prescription together are akin to the trunk of a tree that has superconductivity as one of its branches — as has been shown on the front page of the monograph.

In the next section, we give a brief account of the work done on solar emission lines by following the same approach as was followed for superconductivity. The final section is similarly devoted to quarkonium spectra.

12.2.   Solar Emission Lines

12.2.1.   *Conventional approach to identify solar emission lines*

This approach relies on the experimentally obtained lines in a particular wavelength range, for example, the permitted lines in extreme ultraviolet region of the spectrum. From the measurements of wavelengths of these lines, a theoretical table of terms can be constructed, and by further analysis of this term diagram it is possible to identify transitions which, even though not observed under laboratory conditions, would give lines coinciding with coronal lines. In order to determine more accurately the term differences of various ions, one frequently resorts to extrapolation of the separations of multiplets of isoelectronic series; the uncertainties in the wavelengths of the desired transitions that this procedure may cause need not concern us. What is significant is the fact that after all this is done, one has at command a *huge* set of wavelengths corresponding to transitions between eigenstates of all the known atoms in their various possible stages of ionization.

12.2.2.   *The Rowland puzzle*

Given the above situation, one would imagine the problem of identification of the solar emission lines to be a solved problem: For any observed line, there ought to be a matching line in the set of wavelengths mentioned above. Unfortunately, this hope has not been fulfilled. It was pointed out by Moore[1] in 1956 that many lines in the data comprising about 20,000 lines in the visible spectrum recorded by Rowland 100 years earlier were unidentified. While the accuracy of Rowland's measurements has been confirmed, the challenge of identification of a significant number of lines has not only persisted, but grown with each new datum brought home by the Solar Maximum Mission. Further, the problem is not confined to the visible region of the spectrum. Among the novel ideas put forward to solve the problem was the suggestion[2] that a number of far ultraviolet lines may be attributed to quarks. However this idea was criticized[3] as unlikely.

12.2.3. Hot hydrogen

It is pertinent at this stage to recall the most essential known facts about the Sun. These are as follows: (a) Temperature of the Sun varies from about 10^8 K in the core to about 6000 K at the surface. (b) Its most abundant constituent is hydrogen: For every 10^6 atoms of hydrogen, there are 63,000 of He, 690 of O, 32 of Fe, 3 of Al, 2 of Na, and so on.[4] (c) Coronal temperatures are somewhat in excess of 10^6 K. (d) At the prevalent temperatures, hydrogen in the Sun is predominantly in the form of an electron–proton plasma which is neutral on the whole.

Above facts make one wonder: What would be more natural and attractive than an attempt to explain the bulk of unidentified lines on the basis of one fundamental system — hot hydrogen — consisting of an electron and a proton, Coulomb potential between which is regulated by temperature? This idea is followed up below.

12.2.4. Equation for hot hydrogen

We now seek to incorporate T in the following equation for normal hydrogen valid at $T = 0$:

$$\left(W - \frac{\mathbf{p}^2}{2\mu} \right) \varphi(\mathbf{p}) = -\frac{e^2}{2\pi^2} \int d^3\mathbf{q} \, \frac{\varphi(\mathbf{q})}{(\mathbf{q} - \mathbf{p})^2} \tag{12.1}$$

where $W = E - m_a - m_b < 0$ is the binding energy, E the total energy of the bound pair of an electron (mass m_a) and a proton (mass m_b), e^2 is the fine-structure constant ($\hbar = c = 1$), and μ is the reduced mass,

$$\mu = m_a m_b / (m_a + m_b) \tag{12.2}$$

In order to obtain the desired $T \neq 0$ version of Eq. (12.1) we begin with a BSE like we did for the pairing problem in superconductivity:

$$(\mu_a \gamma_\mu^a P_\mu + \gamma_\mu^a p_\mu - m_a + i\varepsilon)(\mu_b \gamma_\mu^b P_\mu - \gamma_\mu^b p_\mu - m_b + i\varepsilon)\psi(p_\mu)$$

$$= -(2\pi i)^{-1} \int d^4 k I(\mathbf{k} - \mathbf{p})\psi(k_\mu). \tag{12.3}$$

This equation differs from Eq. (2.3) in Chapter 2 in that in involves an electron and a proton rather than two electrons, which cause it to have two different masses. The new symbols in this equation are μ_a and μ_b which are defined as

$$\mu_{a,b} = m_{a,b}/(m_a + m_b). \tag{12.4}$$

Another, *major*, difference between this equation and the one dealt with earlier is that now one cannot adopt MFA, i.e., one cannot assume $I(\mathbf{q} - \mathbf{p})$ to be a constant because pairing is not restricted to a narrow region. $I(\mathbf{q} - \mathbf{p})$ has to correspond to Coulomb potential, i.e., $1/r$, the momentum–space transform of which is

$$I(\mathbf{q} - \mathbf{p}) = -\frac{e^2}{2\pi^2} \frac{1}{(\mathbf{q} - \mathbf{p})^2}. \tag{12.5}$$

Upon carrying out spin-reduction of Eq. (12.2) by multiplying it with $\gamma_4^a \gamma_4^b$, operating with positive-energy projection operators $\Lambda_+^a \Lambda_+^b$, using Matsubara prescription to introduce temperature into the theory — all of which are steps that have been dealt with in detail in Chapter 2, and using Eq. (12.5) for kernel of the equation, we now obtain

$$\left(W - \frac{\mathbf{p}^2}{2\mu}\right) S(\mathbf{p}) = \frac{e^2}{4\pi^2} Q_\beta(W, \mathbf{p}^2) \int d^3q \frac{S(\mathbf{q})}{(\mathbf{q} - \mathbf{p})^2}, \tag{12.6}$$

where

$$Q_\beta(W, \mathbf{p}^2) = \tanh\left(\frac{\beta\mu_a W}{2} - \frac{\beta p^2}{4m_a}\right) + \tanh\left(\frac{\beta\mu_b W}{2} - \frac{\beta p^2}{4m_b}\right). \tag{12.7}$$

Equation (12.6) is the equation we had set out to obtain. It is valid for arbitrary values of T. Note that when $T = 0 (\beta = \infty)$, $Q_\beta = -2$ and hence Eq. (12.6) reduces to Eq. (12.1) as it should.

12.2.5. *Eigenvalue equation for hot hydrogen for high values of T by an application of Fock's method*

Equation (12.6) is a very complicated equation — even when one deals with it in the limit of large values of T (small β). For the latter case we solve the equation by the following sequence of steps.

(a) By making the approximation

$$\tanh(\beta p^2/4m_{a,b}) \simeq \beta p^2/4m_{a,b}, \tag{12.8}$$

we first reduce Eq. (12.7) to

$$Q_\beta(W, \mathbf{p}^2) \simeq (d_1 - d_2 p^2)/(1 - d_3 p^2), \qquad (12.9)$$

where terms of the order of p^4 and higher have been neglected and

$$d_1 = a + b, \quad d_2 = (\beta/4)(1 + ab), \quad d_3 = (\beta/4)(a/m_a + b/m_b)$$
$$a = \tanh(\mu_a \beta W/2), \quad b = \tanh(\mu_b \beta W/2). \qquad (12.10)$$

(b) On substituting Eq. (12.9) into Eq. (12.6), writing S($\mathbf{p}$) as

$$S(\mathbf{p}) = g_l(\rho) Y_l^m(\theta, \eta), \qquad (12.11)$$

multiplying the equation with $Y_l^{m*}(\theta, \eta)$ and then summing over m, we obtain a 1-dimensional equation that involves an associated Legendre function of the second kind.

(c) The equation obtained in the previous step is subjected to the procedure given by Fock[5] in a study that revealed that hydrogen atom has a hidden O-4 symmetry. This is done by using an appropriate representation of the associated Legendre function of the second kind in terms of $P_l(\cos\theta)$ and multiplying the RHS of the equation with $(1/2\pi) \int_0^{2\pi} d\eta'$. These steps lead to the following equation

$$\left(1 - d_3 p_0^2 \rho^2\right)\left(1 + \rho^2\right) g_l(\rho)$$
$$= \frac{\lambda}{2\pi^2}\left(d_1 - d_2 p_0^2 \rho^2\right) \int d\rho' \frac{\rho'^2 g_l(\rho') \sin\theta' P_l(\cos\theta') d\theta' d\eta'}{\rho^2 + \rho'^2 - 2\rho\rho' \cos\theta'}, \qquad (12.12)$$

where

$$p_0^2 = -2\mu W > 0, \quad \rho = |\mathbf{p}|/p_0, \quad \rho' = |\mathbf{q}|/p_0, \quad \lambda = -e^2\mu. \qquad (12.13)$$

(d) Equation (12.12) is transformed into an equivalent equation on the surface of a unit 4-dimensional Euclidean sphere by stereographic projection via Fock variables

$$\rho = \tan(\varphi/2), \quad \rho' = tan(\varphi'/2). \tag{12.14}$$

The equation so obtained is

$$[(1 - d_3 p_0^2) + (1 + d_3 p_0^2) \cos(\varphi)] \sec^4(\varphi/2) g_l[\tan(\varphi/2)]$$

$$= \frac{\lambda}{8\pi^2 p_0}[(d_1 - d_2 p_0^2) + (d_1 + d_2 p_0^2) \cos(\varphi)]$$

$$\times \int d\Omega' \frac{\sec^4(\varphi'/2) g_l[\tan(\varphi'/2) P_l(\theta')}{1 - \cos(\gamma)}, \tag{12.15}$$

where γ is the angle between the directions $(\varphi, 0, 0)$ and $(\varphi', \theta', \eta')$ and $d\Omega'$ is an element of the 4-dimensional solid angle:

$$\cos \gamma = \cos \varphi \cos \varphi' + \sin \varphi \sin \varphi' \cos \theta'$$

$$d\Omega' = \sin^2 \varphi' d\varphi' \sin \theta' d\theta' d\eta'. \tag{12.16}$$

(e) Equation (12.14) is similar to the equation for the hydrogen atom solved by Fock,[5] except for the occurrence of $\cos(\varphi)$ terms on the two sides. Since eigenfunctions in Fock's problem are given in terms of Gegenbauer polynomials as

$$P_{n,l}^{(2)} = [1/(n+1)] \sin^l(\varphi) C_{n-l}^{l+1}(\cos \varphi), \tag{12.17}$$

which satisfy the recurrence relation

$$\cos \varphi P_{n,l}^{(2)} = \frac{(n+2)(n-l+1)}{2((n+1)^2} P_{n+1,l}^{(2)} + \frac{n(n+l+1)}{2(n+1)^2} P_{n-1,l}^{(2)}, \tag{12.18}$$

we adopt the following *Ansatz* for the eigenfunctions of Eq. (12.15):

$$H_l(\varphi) \equiv \sec^4(\varphi/2) g_l[\tan(\varphi/2)]$$

$$= \sum_{k=l}^{\infty} (-1)^k a_k P_{k,l}^{(2)}(\cos \varphi), \ (a_k = 0 \text{ for } k < l). \tag{12.19}$$

(f) Substituting Eqs. (12.19) into (12.15), performing the integration over $d\Omega'$ via the Funk–Hecke[6] theorem and using Eq. (12.18), we are finally led to the following difference equation as the consistency condition for the existence of nontrivial solutions of our problem:

$$b_{k+1} + \frac{2[c_1(k+1) - c_2](k+1)}{[c_3(k+1) - c_4](k+l+1)}b_k + \frac{k-l}{k+l}b_{k-1} = 0, \quad (12.20)$$

where

$$k \geq l, \quad b_{l-1} = 0,$$
$$b_k = \frac{[c_3(k+1) - c_4](k+l+1)}{(k+1)^3}a_k$$
$$c_1 = 1 - d_3 p_0^2, \quad c_2 = (\lambda/2p_0)(d_1 - d_2 p_0^2)$$
$$c_3 = 1 + d_3 p_0^2, \quad c_4 = (\lambda/2p_0)(d_1 + d_2 p_0^2). \quad (12.21)$$

12.2.6. *Solutions of the eigenvalue equation for hot hydrogen for $T \geq 10^5\,K$*

Since Eq. (12.20) is a second-order difference equation, it generally has two solutions: The dominant and the dominated. Its eigenvalues are those values of $p_0^2 = -2\mu\,W$ for which the dominant solution vanishes,[7] and these eigenvalues can be found by the Hill determinant method.[8] The two unknowns in the equation are β (or T) and W. Since the equation was obtained by making the high-temperature approximation, solving it for $T \geq 10^5$ K, we find that it leads to values in the keV range for the ground state ($W_{n=1,l=0}$) of the system. Some examples of such solutions are:

$$W_{1,0}(1 \times 10^5\text{ K}) = 1274.5\text{ eV}, \quad W_{1,0}(5 \times 10^5\text{ K}) = 2862.2\text{ eV},$$
$$W_{1,0}(1 \times 10^6\text{ K}) = 3757.2\text{ eV}, \quad W_{1,0}(5 \times 10^6\text{ K}) = 4996.8\text{ eV},$$
$$W_{1,0}(1 \times 10^7\text{ K}) = 4465.9\text{ eV}, \quad W_{1,0}(5 \times 10^7\text{ K}) = 3194.1\text{ eV}.$$
$$(12.22)$$

12.2.7. *Addressing the SL 304 data*

Equation (12.20) has been employed to address the SL 801, the SL 304, the SL 408A, and SL 408B data. We now give a brief account of how the SL 304 data which pertain to the solar corona and comprise 31 lines between 11 and 22 Å were dealt with.

Table 12.1.   Eigenvalues $|W_{n,l}|$ of Eq. (12.20) for
$1 \le n \le 20$ and $l = 0, 1$, and 2 at $T = 4.27$ K.

		$\mid W_{n,l}\mid$ (eV)	
n	$l = 0$	$l = 1$	$l = 2$
1	5003.65		
2	928.59	933.41	
3	360.24	360.34	360.36
4	197.59	197.60	197.60
5	125.58	125.58	125.59
6	86.99	86.99	86.99
7	63.85	63.85	63.58
8	48.86	48.86	48.86
9	38.59	38.59	38.59
10	31.25	31.25	31.25
11	25.83	21.70	21.70
12	21.70	18.49	18.49
13	18.49	18.49	18.49
14	15.94	15.94	15.94
15	13.89	13.89	13.89
16	12.21	12.21	12.21
17	10.81	10.81	10.81
18	9.64	9.64	9.64
19	8.66	8.66	8.66
20	7.81	7.81	7.81

We first note that the wavelength corresponding to any transition is calculated via

$$\lambda(\text{Å}) = \frac{hc}{\Delta E \text{ (eV)}} = \frac{12398.52}{\Delta E \text{ (eV)}}, \tag{12.23}$$

where $\Delta E = E_1 - E_2$, and E_1 and E_2, the energy eigenvalues of the final and the initial states respectively, must belong to two adjacent columns for the $\Delta l = \pm 1$ rule to be satisfied. Refer now to Table 12.1. It is seen that at $T = 4.2710^6$ K, while the transitions to the state $n = 1$, $l = 0$ from the states $n = 2$, $l = 1$ and $n = 20$, $l = 1$ lead to wavelengths in the range 3.05–2.48 Å, transitions to the state $n = 2$, $l = 0$ from the states $n = 3$, $l = 1$ and $n = 20$, $l = 1$ lead to wavelengths in the range 21.82–13.47 Å. These wavelengths fall in the range of the observed data. As shown in the table, 15 lines in the data are explained at this temperature. For the

Table 12.2.   Observed wavelengths in SL 304 Solar coronal data that can be explained via the indicated transitions between the energy eigenvalues of (12.20) at $T = 4.27$ K.

S. No.	Observed Wavelength (Å)	Theoretical Wavelength (Å)	Identification of Transition $(n_1, l_1 \to n_2, l_2)$
1	21.80	21.82	$3, 1 \to 2, 0$
2	21.60	21.63	$3, 2 \to 2, 1$
3	17.05	16.96	$4, 1 \to 2, 0$
4	16.77	16.85	$4, 2 \to 2, 1$
5	15.45	15.44	$5, 1 \to 2, 0$
6	15.26	15.35	$5, 2 \to 2, 1$
7	14.82	14.73	$6, 1 \to 2, 0$
8	14.40	14.34	$7, 1 \to 2, 0$
9	14.25	14.26	$7, 2 \to 2, 1$
10	14.03	14.02	$8, 2 \to 2, 1$
11	13.82	13.86	$9, 2 \to 2, 1$
12	13.77	13.74	$10, 2 \to 2, 1$
13	13.65	13.66	$11, 2 \to 2, 1$
14	13.55	13.60	$12, 2 \to 2, 1$
15	13.45	13.48	$15, 2 \to 2, 1$

remaining 16 lines four additional temperatures are required, which is not surprising because the film recording the lines was exposed to the entire coronal disc and not to any localized region. As a comparison we note that in the conventional approach eight different atomic spectra are invoked to explain 28 lines, while three lines remain unidentified.

A pertinent question is: How did we arrive at the value of $T = 4.27 \times 10^6$ K in dealing with the above data? We note in this connection that the energy eigenvalues in Eq. (12.22) provide a first guess. In the present instance, we were led to find $W_{n,l}$ for a few values of n and l at $T = (4.0, 4.1, ..4.4) \times 10^6$ K. This was followed by calculation of wavelengths for transitions to the state $n = 1, l = 0$ from the state $n = 2, l = 1$ and to $n = 2, l = 0$ state from the $n = 3, l = 1$ state at each of these temperatures. It was thus found that the wavelength for the latter transition at 4.3×10^6 K was close to the highest observed wavelength in the data. Thereafter it was a minor matter of fine-tuning to arrive at $T = 4.27 \times 10^6$ K, which also led to 14 other lines in agreement with the observed data.

Before proceeding to the next section, we note that the above approach was shown to explain 102 lines in the SL 801 data, comprising 106 lines, by invoking five temperatures: 5.5×10^6, 1.6×10^6, 4.1×10^5, 3.7×10^5, and 2.0×10^5 K. For comparison, we note that in the conventional approach 91 lines were accounted for on the basis of nine species of ions in 31 different stages of ionization.

12.3. Quarkonium Mass Spectra

12.3.1. *Mesons as bound states of quarks*

In the mid-1970s, new mesonic states of matter in the neighbourhood of 3.097 GeV (J/Ψ family) and 9.460 GeV (Υ family) were discovered in the uncluttered environment of e^+e^- annihilations. There followed a great deal of theoretical activity leading to an understanding of the two families in terms of the bound states of $c\bar{c}$ and $b\bar{b}$ respectively, where c (b) denotes the charm (bottom) quark and a bar over it denotes an anti-quark. In this final section we briefly show that the BSE + Matsubara approach followed for superconductivity and the solar emission lines can also be applied to these families by an appropriate choice of the kernel.

Dictated by the fact that free quarks have never been observed, the choice of the kernel for QCD must satisfy the twin requirements of asymptotic free-dom and infra-red slavery. The former of these is a term used to describe *weakening* in the intrinsic strength of the colour force between quarks as they are brought closer together, unlike the electromagnetic force which increases as two charged particles approach each other. The latter term — a corollary of the first — is used to describe *strengthening* of the chro-modynamic force as the quarks are drawn apart. In a series of papers by several authors dealing with the J/Ψ and Y families in the frameworks of both non-relativistic Schrodinger equation and the relativistic BSE, it was shown that the twin requirements of QCD can be met by a variety of poten-tials such as a Coulomb + square root potential, Coulomb + linear poten-tial, Coulomb + harmonic oscillator potential, and a logarithmic potential. We give below an account based on the Coulomb + linear potential (also known as the Cornell potential) in the BSE + Matsubara approach followed by us.

12.3.2.   Kernel of BSE for the $q\bar{q}$ system for the Cornell potential

To deal with the $q\bar{q}$ system we write Eq. (2.1) in Chapter 2 as

$$\chi_p(q) = -\int d^4k \, S_F^{(1)}(p_1) S_F^{(2)}(p_2) G(P, q, k) \chi_p(k), \qquad (12.24)$$

where $S_F^{(1)}(p_1)$, $S_F^{(2)}(p_2)$ are the free quark, anti-quark propagators because of the ladder approximation, p_1 and p_2 are the final 4-momenta of the two quarks of masses m_1 and m_2, p_1' and p_2' their initial 4-momenta; the initial and final relative 4-momenta k and q are, respectively,

$$k = \eta_2 p_1' - \eta_1 p_2', \quad q = \eta_2 p_1 - \eta_1 p_2, \quad [\eta_{1,2} = m_{1,2}/(m_1 + m_2)];$$
$$P = p_1 + p_2 = p_1' + p_2', \quad p_{1,2} = \eta_{1,2} P \pm q, \quad p_{1,2}' = \eta_{1,2} P \pm k.$$

With the above notations,

$$S_F^{(1,2)} = \frac{1}{i(m_{1,2} + i\gamma_\mu p_\mu)} = \frac{m_{1,2} - i\gamma_\mu p_\mu}{i(m_{1,2}^2 + p_{1,2}^2)},$$

and χ taken as the product of free spinors $\bar{u}_{p'\sigma'} u_{p\sigma}$, we have

$$S_F^{(1)}(p_1)\chi_p(k) = \frac{1}{i(m_1^2 + p^2)} \bar{u}_{p'\sigma'}(m_1 - i\gamma_\rho p_\rho)u_{p\sigma},$$

and a similar expression for $S_F^{(2)}(p_1)\chi_p(k)$. Since we wish to identify the observed masses of the Ψ/J or the Υ family with the n $^{2S+1}L_J$ states of the $q\bar{q}$ system, spin-reduction of these expressions is now carried out by Gordon's method which separates out the orbital part from the spin part. On doing so, going over to the center-of-mass system ($P = 0$, iM), and adopting the instantaneous approximation for the $q\bar{q}$ interaction, Eq. (12.24) is obtained as

$$\psi(\mathbf{q}) = -\frac{F_{12}}{(2\pi)^4} \int dk I(\mathbf{q}, \mathbf{k}) J(\mathbf{k})\psi(\mathbf{k}), \qquad (12.25)$$

where $F_{12} = -4/3$ is the "color factor" because the meson must be a color-singlet,

$$I(\mathbf{q}, \mathbf{k}) = [-k_1 M^2 - (\mathbf{q} + \mathbf{k})^2 - 2i\{\sigma_{ij}^{(1)} + \sigma_{ij}^{(2)}\}q_i k_j$$
$$+ \sigma_{ij}^{(1)}\sigma_{il}^{(2)}(\mathbf{q} - \mathbf{k})_j(\mathbf{q} - \mathbf{k})_l - k_2 M(\mathbf{q}^2 - \mathbf{k}^2)x\langle \mathbf{q}|V_{12}|\mathbf{k}\rangle,$$

$$(12.26)$$

$$k_1 = 4\eta_1\eta_2, \quad k_2 = \frac{(m_1^2 + m_2^2)M}{m_1 m_2 (m_1 + m_2)},$$

$$J(\mathbf{k}) = \int_{-\infty}^{\infty} \frac{dk_0}{D(k_0, \mathbf{k})},$$

$$D(k_0, \mathbf{k}) = [(k_0 + \eta_1 M)^2 - E_1^2][(k_0 - \eta_2 M)^2 - E_2^2], \tag{12.27}$$

$$E_{1,2} = \sqrt{\mathbf{k}^2 + m_{1,2}^2},$$

and $\langle \mathbf{q}|V_{12}|\mathbf{k}\rangle$ is the momentum–space representation of the inter-quark potential.

The Cornell potential is

$$V(r) = -\alpha_s/r + \lambda r + C, \quad (\alpha_s, \lambda\text{: coupling constants; C: a constant})$$

which, upon Fourier transformation, leads to

$$\langle \mathbf{q}|V_{12}|\mathbf{k}\rangle = \frac{4\pi\alpha_s}{(\mathbf{q}-\mathbf{k})^2} + \frac{8\pi\lambda}{(\mathbf{q}-\mathbf{k})^4} - \frac{(2\pi)^3\delta^3(\mathbf{q}-\mathbf{k})}{m_1 m_2}. \tag{12.28}$$

It is seen from Eqs. (12.26) and (12.28) that if spin-reduction of the quarks is carried out by Gordon's method, then Cornell potential in momentum space comprises 15 terms. We now recall that when we dealt with the problem of solar emission lines on the basis of Coulomb potential alone, *without* the additional complication of terms obtained via Gordon-reduction, we had to follow a rather formidable mathematical procedure comprising the operation of stereographic projection and application of the Funk–Hecke theorem. It is not immediately obvious as to how such an approach can be implemented in the present case. Hence, after subjecting Eq. (12.27) to the Matsubara prescription and using Eqs. (12.26) and (12.28), we resort to the strategy of transforming Eq. (12.24) into co-ordinate space and numerically solving it by a more familiar method.

12.3.3. *T-generalized equation for quarkonium*

Putting $m_1 = m_2 = m$ (because masses of a quark and its anti-quark are equal), application of the Matsubara prescription to Eq. (12.27) yields

$$J_\beta(\mathbf{k}) = \frac{i\pi \tanh[(\beta/2)(m^2 + \mathbf{k}^2)^{1/2}]}{2(m^2 + \mathbf{k}^2)^{1/2}[M^2/4 - (m^2 + \mathbf{k}^2)]}, \tag{12.29}$$

which, in the limit $T = 0$, reduces to the expression given by Arafah *et al.*[9] On substituting Eqs. (12.26), (12.28), and (12.29) into (12.25) and Fourier transforming, we obtain

$$
\begin{aligned}
(-4M_p r &- (16/3)\lambda' r^2 + (16/3)\alpha'_s - (4/m^2)C'r)u''(r) \\
&= (8/3)(2 + M/m)(\alpha'_s/r + \lambda' r)u'(r) \\
&\quad - [(1/r^2)(S_{12} + 4\mathbf{L.S} + 2M/m - 4(l+1)]u(r) \\
&\quad - (4\lambda'/3)[M^2 r^2 + 2 + 4\mathbf{L.S} + S_{12}/3 + 4\sigma_1.\sigma_2/3 + 4l(l+1)]u(r) \\
&\quad - [C'M^2 r/m^2 + 16Mm^2 r + 4l(l+1)(M_p + C'/m^2)/r]u(r),
\end{aligned}
$$
(12.30)

where

$$
M_p = M(2 + tm\beta), \quad t = \tanh(m\beta),
$$

$$
\lambda' = \lambda w, \quad \alpha'_s = \alpha_s w, \quad C' = Cw, \quad w = M\beta - 2t,
$$

$\psi(\mathbf{r})$, defined as $\int [d^3\mathbf{p}/(2\pi)^3]\exp(i\mathbf{p.r})\psi(\mathbf{p})$, has been put equal to $[u(r)/r]$ times the spin and angular part, and the following relations have been used

$$
\int \frac{d^3\mathbf{p}}{(2\pi)^3}\exp(i\mathbf{p.r})\frac{1}{|\mathbf{p}|^2} = \frac{1}{4\pi r}; \quad \int \frac{d^3\mathbf{p}}{(2\pi)^3}\exp(i\mathbf{p.r})\frac{1}{|\mathbf{p}|^4} = -\frac{r}{8\pi};
$$

$$
\int \frac{d^3\mathbf{p}}{(2\pi)^3}\exp(i\mathbf{p.r})\frac{\mathbf{p}}{|\mathbf{p}|^4} = \frac{i\mathbf{r}}{8\pi r};
$$

$$
\int \frac{d^3\mathbf{p}}{(2\pi)^3}\exp(i\mathbf{p.r})\frac{p_i p_j}{|\mathbf{p}|^4} = \frac{1}{8\pi}\left(\frac{\delta_{ij}}{r} - \frac{x_i x_j}{r^3}\right);
$$

$$
L_i = (-i)(r_j\partial_k - r_k\partial_j), \quad (i, j, k = 1, 2, 3); \quad \mathbf{S} = (1/2)\sigma;
$$

$$
S_{12} = \frac{3(\sigma^{(1)}.\mathbf{r})(\sigma^{(2)}.\mathbf{r})}{r^2} - \sigma^{(1)}.\sigma^{(2)},
$$

$$
\delta^3(\mathbf{r}) = \text{Lim}(r_0 \to 0)\frac{1}{4\pi r_0^2}\frac{1}{r}\exp(-r/r_0) \quad \left(\text{whence } \int d^3\mathbf{r}\delta^3(\mathbf{r}) = 1\right).
$$

12.3.4. *Mass spectra of the* J/Ψ *family*

After incorporating the boundary conditions that ensure proper behavior of the solution at the origin, Eq. (12.30) was solved over an extensive domain

Table 12.3.   Theoretical masses of the $n^{2S+1}L_J$ states of the $b\bar{b}$ family calculated by solving Eq. (12.30) with the parameter values given in Eq. (12.31). Mass of the bottom quark has been taken as Eq. 4.955 GeV.

State	M (expt)	Temperature			
$n^{2S+1}L_J$	GeV	1×10^{12} K	3.165×10^{13} K	1×10^{15} K	3.165×10^{16} K
1^3S_1	9.460	9.465	9.659	9.936	9.914
2^3S_1	10.023	10.022	10.074	9.961	9.916
3^3S_1	10.355	10.341	10.315	9.981	9.918
4^3S_1	10.580	10.597	10.510	9.999	9.919
5^3S_1	10.865	10.825	10.685	10.015	9.921
6^3S_1	11.020	11.037	10.848	10.030	9.922
1^1S_0	9.370	9.389	9.588	9.936	9.914
2^1S_0	9.963	9.990	10.046	9.961	9.916
3^1S_0	10.298	10.318	10.294	9.981	9.918
4^1S_0	10.573	10.578	10.493	9.999	9.919
1^3P_0	9.859	9.004	9.535	9.952	9.915
2^3P_0	10.232	9.972	10.054	9.973	9.917
1^3P_1	9.891	9.852	9.988	9.952	9.915
2^3P_1	10.255	10.194	10.236	9.973	9.917
1^3P_2	9.913	9.920	10.012	9.952	9.915
2^3P_2	10.268	10.248	10.256	9.973	9.917

of parameter space by adopting the fourth-order variable-step Runge–Kutta method. The same set of parameters

$$\alpha_s = 0.6, \quad \lambda = 0.089\,\text{GeV}^2, \quad C = -0.112\,\text{GeV}^3 \qquad (12.31)$$

led to reasonably good fits to the experimental data on both the $b\bar{b}$ and $c\bar{c}$ families; abridged results for only the former of these are given in Table 12.3.

12.4.   Remarks

12.4.1.   Solar emission lines

1. It should be noted that the study of "hot" hydrogen presented in this chapter differs from earlier studies concerned with temperature-generalization of the radiative corrections to the energy levels of the

hydrogen atom. If dynamics of the *basic* system continues to be governed by Eq. (12.1), as in,[10] any such study is merely concerned with the effect of temperature on a system that is assumed at the outset to have $\sim$13.6 eV as its ground-state energy. Such studies have led to the finding that temperature has negligible effect on the spectrum of energy eigenvalues of the hydrogen atom so long as it exists.

2. Even though Eq. (12.6) reduces to Eq. (12.1) when $T = 0$, the spectra of their energy eigenvalues differ significantly. In order to see how this comes about, it is instructive to transform Eq. (12.6) to co-ordinate space. On doing so, as shown in Ref. 11 (b), we are led to

$$\left(W + \frac{1}{2\mu} \nabla^2 \right) u(\mathbf{r})$$
$$= \frac{e^2 D_1}{2r} u(\mathbf{r}) + \frac{e^2 D_2 D_3}{8\pi r} \int d^3 \rho \frac{\exp(-\sqrt{D_2}\rho)}{\rho} u(\mathbf{r} - \boldsymbol{\rho}), \quad (12.32)$$

where

$$D_1 = d_2/d_3, \quad D_2 = -1/d_3, \quad D_3 = d_1 - d_2/d_3.$$

The first term on the RHS of Eq. (12.32) is a pure Coulomb term where the interaction parameter e^2 is modified by the multiplicative factor $D_1/2$, which is always negative and capable of being large owing to its dependence on temperature. This term can therefore mimic the potential in high-Z ions. The second term in Eq. (12.31) is always positive; it involves an integration of a Yukawa-like term over the space of the wavefunction and mimics the screening effect due to the inner electrons. It is hence seen as to why Eq. (12.32) leads to lines which are ordinarily attributed to high-Z ions. We note in passing that mass of the Yukawa particle associated with the second term in Eq. (12.32) is $\sim$1 MeV at 10^6 K. Therefore the short-range interaction generated by this term is not as short as the one between baryons mediated by the pion, whose mass $\sim$140 MeV.

3. The work on the solar emission lines alluded to above has been followed up in a penetrative study by Pande.[11(m)] Salient features of this work are as follows.

   (i) It gives a step-by-step derivation of Eq. (12.6) based on first principles and removes some of the obscurities in the author's earlier derivation.[11(e)]

(ii) Since (i) puts Eq. (12.6) on a firm footing, it inspired the author to explore its further implications. Specifically, both on the basis of Eq. (12.6) and the conventional approach to identify emission lines, he was led to analyse the relatively more recent — and more accurate — data on the emission lines from Solar flares obtained by Phillips *et al.* (*Astrophys. J.* **256**, 774, 1982) and McKenzie *et al.* (*Astrophys. J.* **241**, 409, 1980) and similar data from the coronal active regions McMath 12624 and McMath 12628 obtained by Pye *et al.* (*Mon. Not. Astr. Soc.* **178**, 611, 1977).

(iii) Drawing attention to the fact that flares have been estimated to take place in the upper reaches of the chromosphere at a temperature $\sim$ a few million K, the author has argued that the electro-proton plasma should remain stable at such a temperature for some time. He is then enabled to show that *all* of the 29 lines in the combined data of Phillips *et al.* and McKenzie *et al.* mentioned above can be explained in terms of transitions to the $n = 2$, $l = 1$ state of hot hydrogen at $T = 4.26 \times 10^6$ K from states having different n, l values. As a comparison, the author has pointed out that in the conventional approach, which invokes the ion species of O VII, Ne IX, Fe XVIII, Fe XIX, etc., there remain question marks about the identification of 10 of these lines.

(iv) The author has also shown that lines from the coronal active regions mentioned above can be explained on the basis of Eq. (12.6) with a measure of economy, just as the data on flares were dealt with above.

4. Significantly, Pande[11(m)] has also drawn attention to the fact that a need has been felt in some quarters to supplement the conventional approach to identify solar emission lines because of the First Ionization Potential (FIP) effect which, in the context of emission lines that originate from the photosphere, the coronal active region, and the region of solar flares, he has discussed as below.

5. According to the classical theory of stellar atmospheres, the relative abundances of elements in a star are not expected to vary in its upper layers, unlike in the interior where thermonuclear processes take place. Therefore the relative abundances of elements in the photosphere should be no different from their values in the regions from which the coronal

lines or the flares originate. This has not been found to be so. Analyses of the relative intensities of emission lines from the three regions have revealed that intensities of lines corresponding to low FIP are *anomalously enhanced* in the region where $T \geq 10^6$ K. This implies that elements with low FIP are more abundant in the region where $T \geq 10^6$ K than in the photosphere where $T \approx 6000$ K. This is in conflict with the generally accepted and empirically corroborated picture of uniformity of relative abundances of elements in the upper reaches of the Sun. As Pande has pointed out, hot hydrogen resolves this conflict because it provides an additional channel that yields lines coinciding with those attributed to high-Z ions. It will therefore be seen that the hot hydrogen approach to the identification of emission lines from various regions of the Sun serves to play, at least, a useful supplementary role to the conventional approach.

12.4.2. *Quakonium spectra*

1. As is seen from Eq. (12.29), temperature-generalization of the quarkonium problem leads to an expression that includes the factor $\tanh[(\beta/2)(m^2 + \mathbf{k}^2)^{1/2}]$, which may be approximated as

$$\tanh[(\beta/2)(m^2 + \mathbf{k}^2)^{1/2}] \simeq \tanh(\beta m/2) \qquad (12.33)$$

   by assuming that $m^2 \gg k^2$; this is reasonable because of large mass of the quark ($\sim$4.96 GeV for the b quark).

2. The results in Table 12.3 have been obtained by using Eq. (12.33). It is therefore found that for all values of T for which $\tanh(m/2k_B T)$ remains equal to unity, temperature remains a dormant variable, i.e., it does not affect the spectra of masses given in Table 12.3. For values of T exceeding a certain value, say, T_u, $\tanh(m/2k_B T)$ becomes less than unity and then it begins to play an active role in the dynamics. The notion of mass as a fundamental attribute of a particle is possibly untenable in the region $T > T_u$ where the mesons have variable masses.

3. As temperature is increase beyond T_u, in the limit of high temperatures, the masses of all the bound states in Table 12.3 collapse to twice the mass of a b quark, i.e., the system becomes de-confined because the coupling constants $\to 0$.

212 Superconductivity: A New Approach Based on the Bethe–Salpeter Equation

4. It seems interesting to note that in an alternative treatment[11(j)] of the Cornell potential, one can isolate from the T-generalized kernel the factor $\tanh(\beta k^2/4m)$; if this is approximated as

$$\tanh(\beta \mathbf{k}^2/4m) \simeq \beta \mathbf{k}^2/4m, \qquad (12.34)$$

then T becomes a parameter akin to α_s and λ, see Eq. (12.28). The observed values of different states of the $b\bar{b}$ system are now explained by the following set of parameters:

$$m_b = 4.764 \, \text{GeV}, \quad \alpha_s = 0.078, \quad \lambda = 9.9 \times 10^{-5} \, \text{GeV}^2,$$

$$C = -1.684 \, \text{GeV}^3, \quad T = 2.435 \times 10^{11} \, \text{K}. \qquad (12.35)$$

A significant difference between this scenario and the one dealt with earlier is that de-confinement now results in the limit of $T \to \infty$, i.e., quarks are permanently confined.

Notes and References

1. C.E. Moore, *Vistas Astr.* **2**, 1209 (1956).
2. O. Sinanoglu, B. Skutnik and R. Tousey, *Phys. Rev. Lett.* **17**, 785 (1966).
3. W.R. Bennett, Jr., *Phys. Rev. Lett.* **17**, 1196 (1966).
4. D.M. Considine, ed., *Scientific Encyclopedia* (Van Nostrand Reinhold, NY, 1989).
5. V. Fock, *Zs. Phys.* **98**, 145 (1935).
6. A. Erdelyi, ed., *Higher Transcendental Functions*, Vol. 2 (Bateman Manuscript Project) (McGraw Hill, NY, 1953).
7. A. Hautot, *Phys. Rev. D* **33**, 437 (1986); M. Znojil, *Phys. Rev. D* **34**, 224 (1986).
8. S.N. Biswas *et al.*, *Phys. Rev. D* **4**, 3617 (1971); — *J. Math. Phys.* **14**, 1190 (1973); J.F. Kilinbeck, *Microcomputer Quantum Mechanics* (Adam Hilger, Bristol, 1983).
9. M. Arafah, R. Bhandari and B. Ram, *Lett. Nuovo. Cim.* **38**, 305 (1983).
10. B.Y. Cha and J.Y. Yee, *Phys. Rev. D* **32**, 1038 (1985).
11. This chapter gives a very brief account of the matter that is covered in far greater detail in the papers noted below.
   (a) G.P. Malik and L.K. Pande, *Phys. Rev. D* **37**, 3742 (1988); dealt with T-generalization of the Wick-Cutkosky model which led to T-dependent discrete energy spectrum and suggested that it should be interesting to similarly T-generalize the more realistic Coulomb potential.
   (b) G.P. Malik and L.K. Pande, *Pramana, J. Phys.* **32**, L 89 (1989); showed that generalization of the Coulomb potential between an electron and a proton in a *medium at a non-zero temperature* leads, in the high-T limit, to energy eigenvalues in the keV range — suggesting a possible application to solar emission lines from solar corona.

(c) G.P. Malik, L.K. Pande and V.S. Varma, *Astrophys. J.* **379**, 788 (1991); addressed the solar emission lines in SL 801, SL 304, SL 408 A and 408 B data on the basis of the equation obtained in (b).

(d) G.P. Malik, U. Malik and V.S. Varma, *Astrophys. J.* **371**, 418 (1991); the need to deal with intensities of the lines in (c) above led us to present in this paper a new method for the calculation of radial matrix integrals for electric dipole transitions in hydrogen (it has been called the method of Malik, Malik, and Varma by James K.G. Watson, FRS, in *J. Phys. B: At. Mol. Opt. Phys.* **39**, 1889, 2006). Formulae obtained in this paper, which is cited in the *Astrophysical Formulae* by K.R. Lang, are alternatives to Gordon's classic formulae (W. Gordon, *Ann. Phys.* **2**, 1031, 1929); they are particularly useful for transitions between neighbouring states characterized by large values of the principal quantum numbers, and have also been used for shock waves.

(e) L.K. Pande, *Pramana, J. Phys.* **37**, 39 (1991); without employing the short-cut of the Matsubara recipe, this paper gives *ab initio* derivation of Eq. (12.6) based completely on the formidable apparatus of FTFT.

(f) G.P. Malik, U. Malik and V.S. Varma, *Astrophys. and Space Science* **199**, 299 (1993); dealt with certain conceptual issues of the approach presented in (d) and a predictive test of it.

(g) G.P. Malik, U. Malik, L.K. Pande and V.S. Varma, *Astrophys. J.* **371**, 447 (1995); dealt with calculation of the relative intensities of the SL 304 data.

(h) G.P. Malik, R.K. Jha and V.S. Varma, *Astrophys. J.* **503**, 446 (1998); in this paper Eq. (12.6) is transformed to co-ordinate space and shown — via confluent hypergeometric functions — to lead to Eq. (12.20) without employing the rather specialized techniques of stereographic projection and the Funk–Hecke theorem that were used in the momentum–space treatment of the equation.

(i) L.K. Pande, *Proc. Nat. Acad. Sci. India*, **71 (A)**, IV, 339 (2001); this independent study carried out the same program as in (h) by employing associated Laguerre polynomials.

(j) R.K. Jha, G.P. Malik and V.S. Varma, *Astrophys. and Space Science* **249**, 151 (1997); employing the BSE + Matsubara framework with a Coulomb plus a linear kernel, in this paper the mass spectra of the J/Ψ and Y families were obtained via the bound states of appropriate quark-antiquark systems.

(k) G.P. Malik, R.K. Jha and V.S. Varma, *Eur. Phys. J. A* **2**, 105 (1998); in this paper a more detailed study of the J/Ψ and Y families was carried out with a Coulomb plus a linear kernel.

(l) G.P. Malik, R.K. Jha and V.S. Varma, *Eur. Phys. J. A* **3**, 373 (1998); quarkonium spectra were studied with a logarithmic, and a Coulomb plus square-root kernels.

(m) L.K. Pande, *J. Mod. Phys.* **7**, 25 (2016).

Summing Up

It is prudent to ask: How exactly does this monograph advance our knowledge about superconductivity?

If it should be asked whether it enables one to predict the ratios in which different elements ought to be combined so as to form a superconducting compound with a given T_c, then the answer is no. We should like to note however that such a prediction cannot be made even for *non-superconducting* compounds since, for example, given all the properties of hydrogen and oxygen, one cannot *a priori* predict that H_2O should have the properties that it has.

If — and there can hardly be two opinions about it — BCS theory advanced our knowledge about superconductivity, then the generalization or extension of it presented in this monograph is a further similar step. This is so because it brings the understanding of (non-elemental) HTSCs at par with that of elemental SCs in the original theory. Also suggested herein is, for the first time, an explanation of the Meissner effect based on the *dynamical* equation for pairing. But there is more than this here.

At the heart of this monograph is demonstration of the equality between Δ and $|W|$ which has been implied in the literature all along, but not proved. While this result is important *per se*, and some of its consequences — like dynamics-based equations for H_c and j_c of both elemental and non-elemental SCs — were explored here, it opens up additional avenues for further work. Among these is the suggestion that all HTSCs in general, and those that exhibit a multitude of gaps in particular (e.g., iron-pnictide SCs), be addressed via equations incorporating chemical potential(s), just as La_2CuO_4 was addressed. While usefulness of such a *novel* approach was demonstrated in the context of $SrTiO_3$ and HFSCs too, we believe that a lot more needs to be done to find concrete clues about raising T_c. We draw attention in this connection to the increase in T_c when Tl-2212 is obtained from Bi-2212 by replacing Bi by Tl and Sr by Ba. In a discussion in Chapter 4, it was pointed out that theory needs to take into account the changes in the values of E_F (or μ) and the other constituents of λ when these substitutions are made. Since μ is directly related with the number density of charge carriers, this is a feature that is amenable to experimental verification via, e.g., the Hall Effect.

Not addressed in this monograph is the p- or d-wave superconductivity. The equality between Δ and $|W|$ suggests the possibility of doing so via a T-generalized BSE. The kernel of this equation can be taken to comprise the usual BCS model interaction and a perturbation term corresponding to the p- or d-wave interaction. Conceptually simpler than the variational approach followed in the original BCS theory and easier to implement, such an approach would parallel one that is followed for the $\ell \neq 0$ states of the hydrogen atom — except that it will have T as an additional parameter.

A key feature of our approach is the characterization of a non-elemental SC by MDTs. This is a feature that has been vindicated by experiment, as discussed in Chapter 4 and in Appendix 11A.

It was noted at several places in the monograph (e.g., Chapters 3, 5, 6, 10, and 11) that, strictly speaking, the interaction parameters (λs) in the theory should be regarded as T-dependent. The need to do so is indicated even at the level theory deals with elemental SCs since, as is seen from Tables 3.1 and 3.2, the value of λ for any element calculated via the equation for T_c is invariably different from its value calculated via the equation for Δ_0. Because elements have rather low T_cs, it turns out that for most such SCs, $\lambda(T_c) \approx \lambda(\Delta_0)$; these are the "good" SCs for which $2\Delta_0/k_BT_c \simeq 3.53$. This relation is not satisfied by some SCs, such as Nb and Hg, because $\lambda(T_c)$ for each of them differs considerably from $\lambda(\Delta_0)$; hence coinage of the term "bad actor" SCs for them.

Just as there are "good" and "bad" elemental SCs, there may well be such HTSCs. To see this, we recall that each of the λs for the latter has the following structure:

$$\lambda \equiv [N(0)V] \approx E_F^{1/2}V.$$

Therefore, in going from T_c to 0 K, it can happen that the change in V is offset by a change in E_F so that, effectively, λ has the same value at both these temperatures. It seems to us that this happens, e.g., for MgB_2 because the values of its T_c and Δs could be explained with only a minor fine-tuning. But this is serendipitous. One must not expect this to happen in general, particularly because we are seeking a theory in which T_c exceeds 0 K by two orders of magnitude. We note that confirmation of the idea that interaction parameter(s) for SCs are temperature-dependent has been provided by the recent experimental work of Mansart *et al. Phys. Rev. B* **88**, 054507 (2013) — which was briefly discussed in Chapter 5.

To conclude, it seems to us at this stage that (a) there is no alternative to understanding superconductivity other than via formation of CPs, (b) the generalization or extension of BCS theory for elemental SCs by incorporating into it the concepts of MDTs and formation of CPs via multi-phonon exchange mechanism provides a robust platform to deal with HTSCs and that, while doing so, we need to (c) incorporate chemical potentials into the equations for not only their T_cs and Δs, but also H_cs and j_cs, and (d) allow for all the parameters of the theory, including the θs, to be temperature-dependent.

We believe that BSE provides an appropriate framework for realization of the above program.

Acknowledgments

The author is indebted to his father's sister, Dr. K. Bahl (nee Kunti Malik), for taking him under her wings[1] in the formative years before class X and inspiring him affectionately to achieve the high standards set by her.

He remembers with gratitude another sister of his father, Kaushalya Devi and her husband Dr. Ram Murti, for the same reasons as above; G.P. Bhatt, D.V. Falls, and E.P. Zachariya, his teachers during the years up to class X at A.P. Mission (now Church of India) School, Dehra Dun, India, for the life-long influence of their nurturing; R.C. Majumdar and D.S. Kothari, who inspired him like they did about three generations of students at the Physics Department, University of Delhi (UD); S.N. Biswas, his PhD supervisor at UD, from whom he learnt much of the *technology* that he has used in his work, and for sane advice to work on off-beat ideas *after* obtaining a PhD, A. Salam from whom, as a post-doc at ICTP, Trieste, he imbibed the *essence* of doing research in Physics.

He is grateful to A.N. Mitra, his quantum field theory teacher at UD, for generous encouragement; L.N. Cooper for apprising him about papers that opened further avenues for work and for encouragement, and D.C. Mattis for always promptly responding to his queries, for encouragement and for several invaluable suggestions for improvement of this monograph.

He thanks his collaborators, L.K. Pande, V.S. Varma, Usha Malik (his wife), R.K. Jha, Akhilananda Kumar, M. de Llano, and I. Chávez, for patiently bearing with him in tackling problems that were, more often than not, off-beat and which required many rounds of correspondence before acceptance.

He acknowledges valuable correspondence with V.Z. Kresin, who stimulated his interest in the thermal conductivity of superconductors; M. de Llano, who stimulated his interest in the BCS-BEC crossover physics, M. Fortes in connection with Chapter 3, and D. Eagles, who stimulated his interest in $SrTiO_3$.

---

[1] This was necessitated by our migration to Delhi from Peshawar where my father taught Chemistry at King Edwards College and we had a well-settled life, prior to the turmoil caused by partitioning of India into two nations.

He is grateful to the Organizers of EUROQUAM 2010 (Ischgl, Austria), QFS 2010 (Grenoble, France), QFS 2012 (Lancaster, UK), and QCM 2014 (Obergurgl, Austria) for financial support enabling his participation, and the Organizers of New^{3}SC-8 2011 (Chonqing, China) for an award in recognition of his work; the Editors of The *Astrophysical Journal, International Journal of Modern Physics B, Physica B, Physica C,* and *Physical Review D,* among others, for painstakingly ensuring that a fair verdict was reached for his submissions even when there was a negative report by one of the reviewers who were approached — ten of them in one case.

He takes this opportunity to acknowledge that in devoting himself to the work that has led to this monograph, he had aspired to emulate Ashok Goyal (his friend and colleague at UD) and R. Rajaraman (his senior at UD, friend and later a colleague at JNU); former for the sustained work that he has done over more than three decades and latter for the work that he did in a single year (1975) while he was at Princeton.

Last but not least, he considers it a privilege to have been associated with Jawaharlal Nehru University for about thirty years for the *freedom* and *leisure* it allowed him to pursue his interests; for the same reasons, he is indebted to ICTP, Trieste, and The Inter-University Centre for Astronomy and Astrophysics, Pune, which he was privileged to visit several times for varying periods of time.

Index

Printed in the United States
By Bookmasters